HIGH VALUE, LOW COST

HIGH VALUE, LOW COST

How to Create Profitable Customer Delight

BRIAN PLOWMAN

FINANCIAL TIMES
PITMAN PUBLISHING

PITMAN PUBLISHING
128 Long Acre, London WC2E 9AN

A Division of Longman Group Limited

First published in Great Britain 1994

British Library Cataloguing in Publication Data
A CIP catalogue record for this book can be obtained from the British Library.

ISBN 0 273 60437 6

10 9 8 7 6 5 4 3 2 1

Typeset by Northern Phototypesetting Co Ltd., Bolton
Printed and bound in Great Britain by Biddles Ltd, Guildford and King's Lynn

The Publishers' policy is to use paper manufactured from sustainable forests

CONTENTS

FOREWORD

In most sectors of any country's economy, the battle to sustain and increase corporate profitability grows ever more arduous. Margins are caught in a pincer movement by, on the one hand, the steady improvement of the competition, and on the other, the increasing commercial awareness of customers.

Competitors can't be prevented from improving, but the way customers behave and work creates an opportunity. It is an opportunity that can be addressed through the application of Business Process Management (BPM). BPM goes beyond just re-engineering current processes to meet current customer and business needs. BPM is about transforming the business, integrating the external factors that influence the business, its positioning, and aligning the internal factors within the business, its capability.

Many companies have a vision, often expressed as a simple phrase to unite employees and point the way forward to a better world. But is such a simple statement enough to polarise the whole company into action? In this book, the concept of a differentiating customer proposition is developed which provides a unifying theme, the strategy even, for the whole company. Emerging from recession, the world has changed. By creating a differentiating customer proposition and achieving it through the implementation of BPM, companies can meet the challenges the new world circumstances create on two fronts:

The customer: Customers are becoming more sophisticated and better informed. People and businesses demand value for money in *their terms*, by *their criteria*, with products and services conveniently available at places *they choose*. Failure to achieve this and customers look elsewhere.

The organisation: To remain price competitive, and be able to innovate and grow the business, companies need to gravitate towards becoming the *lowest unit cost provider*. They must break through the constraints of traditional practices, attitudes and beliefs to become *competitively unassailable*, or they risk sinking without trace.

There is no room for complacency. We need to grasp every opportunity not only to be effective, but also to be truly competitive, winning more market share by adding value to the customer without mortgaging future investment through cut-throat price wars. We need to increase the total size of our chosen markets to create and raise prosperity for all.

This book is not a panacea; life isn't that simple. However, the book provides a route map to help all those who want to follow the steps on the BPM journey towards corporate transformation, and, as such, is an absorbing read for people who wish to improve their navigation and avoid the pitfalls along the way.

With this book I wish you a safe and pleasant journey on the road to corporate transformation.

MAURICE HUNT
Deputy Director General
The Confederation of British Industry

ACKNOWLEDGEMENTS

For many years Myron Tribus has been an articulate and insightful speaker at many Deming conferences around the world. I am indebted to him for his generous permission to relate and embellish his 'Germ theory of management' which is the inspiration for the 'Doctor's dilemma' story in chapter 1.

The section 'Branding and brand positioning' in chapter 4, is largely based on a contribution from John Hegarty and Gerry Alcock, both of whom are directors of The Brand Positioning Consultancy, based in London, UK. John and Gerry are active in this field and I am grateful for their permission to relate the issues concerning branding and corporate positioning, and for their permission to outline their methodology for creating a brand. For further details, they can be contacted on (UK) Tel: 071-240 8691, Fax: 071-836 3632.

The section 'Image and attitudinal tracking' in chapter 4 is largely based on a contribution written by David Lowings, a director of Dragon International, based in London, UK. The company is active in product development and branding based upon the understanding of consumer-relevant trends. For further details, they can be contacted on (UK) Tel: 071-229 6090, Fax: 071-229 4014.

The section 'Distribution and channel modelling' in chapter 4 is largely based on a contribution from Professor Martin Clarke, Managing Director, and Jonathan Walker, Business Manager, of GMAP Ltd, Leeds, UK. The GMAP modelling techniques outlined in this section have been developed over the past twenty years in the School of Geography at the University of Leeds. GMAP have developed the approach to its full commercial potential in many sectors. For further details, they can be contacted on (UK) Tel: 0532 446164, Fax: 0532 343173.

This book pulls together the shared experiences of all the people at Develin & Partners, all of whom have provided anecdotes and examples from their own experience working in many sectors on a range of projects. Develin & Partners is based in London, UK, and is active in the fields of Business Process Management, Activity Based Cost Management, Total Quality Management, and Business Led IT Strategy. For further details, they can be contacted on (UK) Tel: 071-917 9988, Fax: 071-895 1357.

Finally, I am also grateful to the directors, managers and staff of all the client companies with whom Develin & Partners has worked. It is through working with our clients that our thinking on Business Process Management has developed and the real world provides the reality by which our thoughts are judged. I am also grateful to everyone at Develin & Partners who constantly act as a test model on the basis that we try only to preach that which we have first practised on ourselves.

To Maureen, Tracey, Tanya and Sophie,
for their patience during the lost weekends.

PREFACE

I believe that people want to do a good job and take pride in the work that they do. When left to their own devices, they will naturally search for a better way to do things. Not to save time so they can do nothing, but for the basic pleasure of seeing a job well done and the pleasure of seeing a better way of doing things. Naturally, they will help each other, as to provide help on one occasion is to receive help on another. Where people understand the risks and consequences of their actions they will behave responsibly and not take action detrimental to others and the business. When given the proper tools to do a good job, and they are competent to do something then they can be left to do it. If we expect more from people, then with additional tools and help to grow their competency, they will do the job expected of them. So why is it that in the work environment, so few people feel the freedom to really enjoy work? Maybe the problem starts at birth.

As infants, we spend our early years searching for answers to the question, 'Who am I?'. We have to make a niche for ourselves in a crowded world of siblings, parents and friends. We learn attention-seeking devices and reward strategies, and the use and abuse of power among our siblings and young friends. At school, we spend some time searching for answers to the questions, 'How does that work?', and 'Why does that happen?'. But we also learn that the teaching process re-inforces the attention-seeking devices, reward strategies and the use and abuse of power. We learn that if we are taught badly, despite every effort to learn, we are rewarded with low marks.

At work, we start by concentrating on how things are done as the minimum requirement to do the job and stay in the job. We follow the rules of the game to get promotion. Just like the early years, we learn attention-seeking devices and reward strategies, and the use and abuse of power in the working environment. We receive promotions and with each promotion we receive re-enforcement that we display the right behaviour and that our knowledge of managing is adequate. We learn how to be noticed by competing with our peers, even where this is in conflict with generating higher overall business performance by working together. The rules tell us that our staff should not be trusted and we make them suffer continuous audit of their decisions. We learn that their performance should be tracked by measuring things that can be measured simply, despite their irrelevance to making the business more effective. We learn that they should be stretched and asked to do things beyond their competence and we make them personally accountable so they suffer the consequences of failure.

A few lucky ones reach the top of the business and that confirms that they have one hundred per cent of the knowledge they need to run the business. By definition, those below have less knowledge. At the top, you can listen patiently to the ideas of those beneath you but you can also draw comfort from knowing that you know more, and, with your authority, can squash anyone taking action contrary to your opinions. With such complacency the conditions are set to make victims of the staff and victims of the customers.

So why did the work environment ever get to be like this?

Chaos theory teaches us that a chaotic system exists when the output is not predictable but is very sensitive to small changes in the starting conditions. A butterfly flaps its wings in Australia and creates a hurricane in the North Atlantic. Maybe the starting point for setting up the first company was like this. The conditions we find in businesses today are the consequences of one unfortunate starting point that led to functional parochialism and a belief in hierarchical importance. But the final outcome we see today was never predictable. We cannot turn the clocks back, but we can create another starting point. We do not have to be hostages to this misfortune.

In 1988, myself, Max Hand, David Baines, Robin Bellis-Jones and Nick Develin found we could not turn the clocks back so we created another starting point. Like me, they also believed in people. In Develin & Partners, we try to turn the old way of doing things on its head. We trust people, we encourage and assist their growth in their personal competences, we do not structure them in conflict with each other or the needs of our customers. We try to make every day one that is a pleasure to live through. It is hard. Every day we still make mistakes; our past is so ingrained, the old rules and behaviours are etched into our brains. We have much to learn.

We are indebted to our staff and our clients for having the patience to listen to our views, show us the error of our ways and make us better human beings. For us, corporate transformation meant starting our own company so we could practise what we preach. Many others will not have this good fortune. They will have the extraordinarily difficult task of attempting corporate transformation by challenging everything in their businesses against the tide of history.

Current business methods are chaotic systems, and many current management practices will still give unpredictable results. Companies should draw little comfort from perpetuating the old ways and staying as they are. That butterfly has a lot to answer for.

1

RECOGNISING THE NEED FOR CHANGE

The researches of many commentators have already thrown much darkness on this subject, and it is probable that, if they continue, we shall soon know nothing at all about it.

Mark Twain

INTRODUCTION

'Business Process Management' (BPM) might be regarded by cynics as yet another grand title for applied common sense. But organisations which address BPM are recognising three fundamentally important issues:

1. Just meeting their customers' current needs does not provide the strategic long-term **positioning** of the business that ensures that it will be competitively and financially successful in local and world markets in the future.
2. Traditional methods of improving the business only focus on separate functional issues and individual internal customer needs within the current processes.
3. The totality of all the internal processes, the business's **capability**, needs to be constantly driven by the need to create the future positioning of the business.

Conventionally, positioning alone has been seen as a largely separate set of functional tasks split within the marketing, strategic planning and finance functions. Capability changes are generally the preserve of individual operations functions which could be working to an agenda determined by local measures of productivity. Understanding and addressing the inextricable linkage between positioning the business and having the appropriate capability is itself the key process in the organisation. Linking positioning and capability is thus the **job** of the executive and senior management in leading the business to a better future. BPM is focused to help the executive do this job better.

A fundamental problem

Internally, the delivery of products and services to customers is the end result of co-ordinated activities by different groups working within the business. Companies seem to go through an irreversible life cycle that leads them towards specialisation, complexity and functional parochialism. However hard we try to avoid the situation developing, the entrepreneurial start of the business, where everyone can virtually do any of the necessary tasks within it, slowly evolves into functions. As functions clone staff together over the years, the rigid development of formal functional structures has provided the opportunity for functions to become fortresses, the contents of which become the jealously guarded property of the occupants. Inside each fortress allegiances are high and people speak their own language, a mechanism to spot intruders and confuse communication.

The natural variation of any process leads to errors. In the functional fortress, it is easy to blame others rather than cross the functional boundaries and resolve multi-functional problems. For companies that are multi-divisional and international, not knowing who to blame is a source of internal frustration. For many members of staff and management, this is their working life. Even moving from one company to another rarely seems to bring us to a different working environment. If we cannot conceive of something better then we accept the situation as simply being business-as-usual and do the best we can in the circumstances.

As shown in Figure 1.1, the fundamental problem for many companies lies in their structure. Organisations are hierarchical, while the transactions and work-flows that provide service and products to customers remain, as always, a horizontal path through the business. The traditional management structure causes managers to put functional needs above those of the multi-functional processes to which their departments contribute. This results in departments competing for resources and blaming one another for the company's inexplicable and continuing failure to meet or exceed current customers' needs efficiently, as well as an inability to see how to put in place a series of multi-functional processes, focused on the customer, that will provide future competitive differentiation.

Quite subtly, the structure determines who really is the customer. It is the person you work for, the person who directs your activities, the person who appraises your performance. Hierarchical structures have taught us to know that we do as we are told by those above us, even where this is in conflict with meeting the needs of the source of our revenue.

People in organisations can sense that they are in a learning environment. They learn how to get on by perpetuating the role models set by their managers and directors. They learn how to be noticed by competing with their peers, even where this is in conflict with generating higher overall business performance by working together in teams. They learn that they cannot be trusted and suffer continuous audit of their decisions. They learn that their performance is tracked by measuring things that can be measured simply, despite their irrelevance to making the business more effective. They learn that they will be asked to do things beyond their competence and will personally suffer the conse-

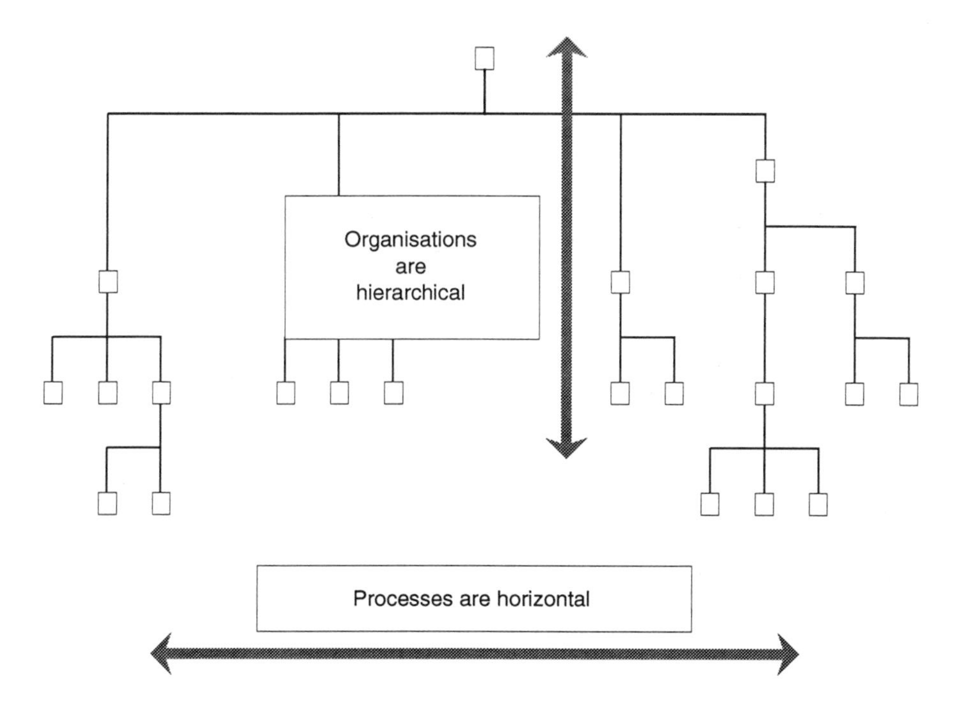

Fig. 1.1 Processes and hierarchical conflict

quences of failure. They learn that to challenge the long and accepted norms of the structure and way of running the business is heretical and can be a significant career-blocking factor.

We should ask ourselves, if there is a better way what would happen if our competitors found it first? Would we respond at all? Would we catch them up? Would we overtake them? Good companies constantly search for better ways and get ever better as a result. Poor companies close their corporate minds and hope for the best, comforted by the old ways that have served them well in the past. In the 1990s and beyond, complacency will be the poison that will eventually finish off many companies.

A way forward

Business Process Management addresses all the necessary steps to undertake, to create a cost-effective business in the short term, and a truly competitively differentiated business in the longer term. This book promotes the simple but powerful insight that links positioning and capability as the mechanism for designing and delivering the vision, and BPM as the overall process to achieve it. BPM is a journey with a distinct path and a specific set of actions along the way; BPM will be the mechanism:

- to create the long-term future positioning of the business and its future capability;
- to create short-term cost effectiveness and improvement to current customer service;
- to initiate continuous improvement from the base of the current, but improved, processes;
- to introduce a knowledge of product and customer profitability;
- to re-engineer the business radically and provide clear future competitive differentiation;
- to address the cultural barriers that prevent effective cross-functional and hierarchical working;
- to introduce leadership and a role for managers and empowered staff.

BPM is not a panacea or a short cut. It demands time and energy. It demands an open mind and a holistic view of your business in relation to the business environment in which the company is operating. It demands a constant challenge of the accepted norms. It demands a commitment to stay on the journey, despite the pitfalls and barriers that confront every step along the way. Above all else, it demands that people should enjoy being at work as part of enjoying life: life is too short to expect anything else.

A new perspective on the business

In Figure 1.2, we can map any business on two axes. BPM considers the *positioning* of the business; those external environmental and customer factors that influence where it wants to be in the future to ensure a growing revenue, and then changing its *capability* to make the journey, by fundamentally challenging and re-engineering the multi-functional processes within the business. BPM maps how processes work in practice and where the problems and failures occur, and provides the base from which to re-engineer the company's capability from small-scale improvements to innovative re-design. BPM goes to the heart of delivering customer service efficiently by addressing not only the mechanisms of doing so, but also tackling the cultural barriers that so often impede progress.

Positioning

Actions to determine positioning draw on a range of techniques. These must have a market focus but often, for many businesses, the internal business objectives still predominate and actions, such as just cost-cutting, are put in place to redress financial performance. Positioning starts with the customer and the competitive environment in which the company is operating. If you get it right for the customer, relative to your competitors, then the financial performance will follow.

Developing the future positioning of the business is a science. For many

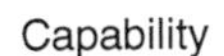

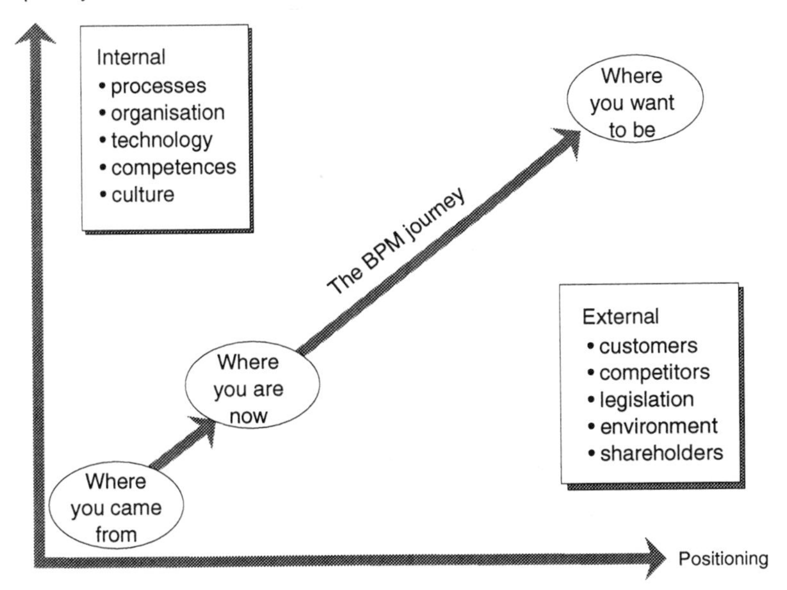

Fig. 1.2 Positioning and capability are linked

businesses, this activity is a poorly co-ordinated mix of outputs from the marketing, strategic planning and finance departments, the whole often usurped by a range of parochial functional plans from the operations departments. At worst, the managing director's or chairman's whims are the sole creators of business direction. The key drivers need to be those factors that are critical for the success of the business, and it is vital that these factors are written in the customers' language in order to create a differentiating customer proposition that locks out the competition.

Capability

The change in capability is driven by positioning. Key processes are designed to deliver each critical success factor. Measures will need to be focused on determining that all the processes are successfully delivering.

Starting with radical and innovatory re-engineering to change the capability fundamentally runs the risk of starting the changes from a base of current business ineffectiveness. In many instances, BPM will need to start with a clean-up of the current processes. Not only will this provide short-term cost savings to be used for improvements to current levels of customer service, but it will also create a vehicle for employee involvement in challenging and changing their own working environment.

Overcoming the barriers of functional parochialism prepares the organisation for both the challenge of radical re-engineering of the business, and the necessary changes to attitudes, beliefs and behaviours within management that are required to ensure that there will be no slip back to the old ways. Managers have a key role in the transformation of an organisation through BPM. By working to help their staff, the victims of poor processes will be freed to add their knowledge to creating continuous improvement at any point on the spectrum of changed capability.

The business strategy is therefore the management of the inextricable link between positioning and capability, and is itself a key process. The future position and the capability to get there is the delivery of the vision. This process is owned by the excutive team. It is their key responsibility. It cannot be delegated.

Making it happen

As shown in Figure 1.3, there are four key elements to achieving a successful ongoing implementation of Business Process Management.

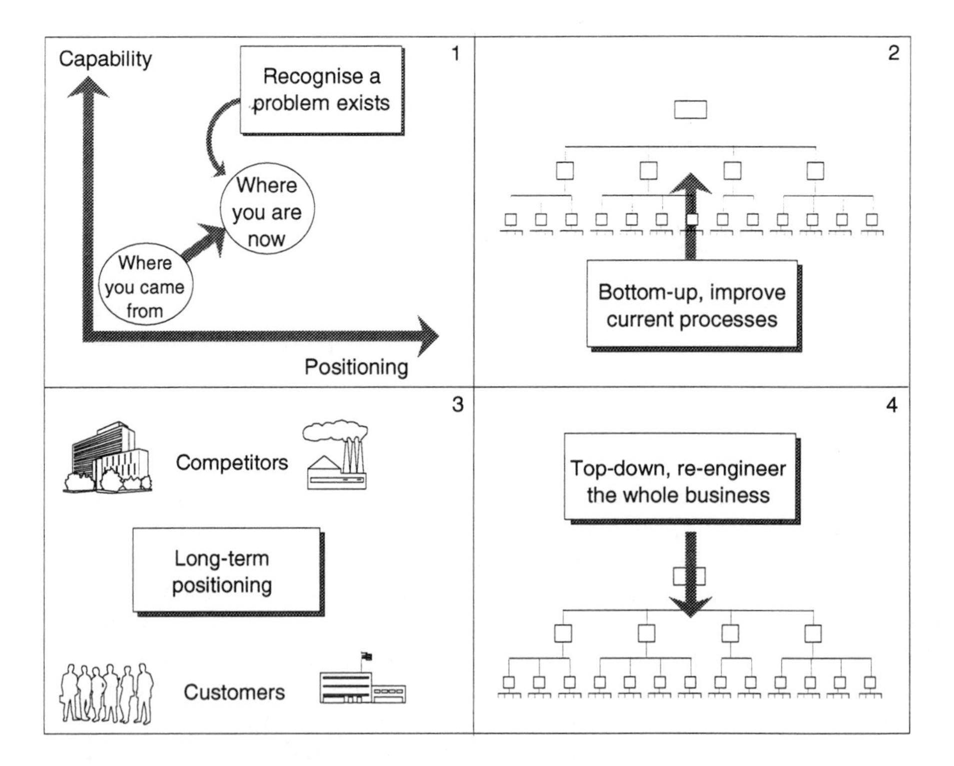

Fig. 1.3 The four key elements of Business Process Management

1. First, the company needs to recognise that it has a problem that requires the level of change that BPM entails. By viewing the business on the axis of positioning and seeing the relationship to the axis of capability, the company will begin to see how customer and business issues can be brought together to impact on realigning the business processes.
2. Through a bottom-up process, staff and managers measure and analyse the current capability. From this base, staff and managers propose short-term options for change which are brought together through forums of internal customers and suppliers. This stage improves the current business, puts in place mechanisms for continuous improvement, and creates the foundation for fundamental re-engineering.
3. From customer and competitor research, the key positioning conclusions create the vision, expressed in terms that will delight customers and turn them into advocates of the business, and point the direction to innovative changes to capability. Through a process of deployment, every employee sees the same vision, but expressed in a language they understand, and thus become aware of how their own element of the process can contribute to making the vision a reality.
4. Using the vision as the driver, a top-down approach re-engineers the key business processes, starting with the customer-facing issues and building back into the business. The need is to create a proposition for customers that leads to retention and advocacy, and to deliver this through an internal framework that ensures lowest unit cost. The innovative approach also completes the process of transforming management attitudes and behaviour.

THE ISSUES FACING ORGANISATIONS

Nothing stands still. World competition ebbs and flows across different parts of the globe. Customers' needs and perceptions change. We carry the cultural baggage of ways of working together that may have worked in the past but lack credibility today. We squander the little knowledge we learn and fall back on old trusted techniques. Every day we look for some stability, some firm ground from which to plan our own future without needing to take account of anyone else. If we believe that this state will ever come about then we are living in a state of self-delusion. The outside world can be a cruel place.

World competition

In which ever year we stop to think about it, we seem to feel that things must get better. Surely a period of growth and stability is just around the corner. But when we really think about it, and look back in time, we see varying degrees of turmoil on a national and international scale.

There are people alive today who have lived through most of the twentieth century. They have seen a world at war a number of times, with allegiances changing during and between each conflict. There are teenagers who have seen traumatic change in Eastern Europe and think nothing of playing arcade-style games on their home TVs, with all the high-technology hardware in their hands coming from the Far East. More recently, there are people at any level in an organisation who have changed their view of the future from one of long-term planning through to retirement, to one of hoping they still have a job the following week. In the developing world, throughout any timescale, people have struggled to meet the daily needs of survival.

The bottom line of human endeavour, in life and in business, is that survival is not compulsory and the comfort of the *status quo*, when we are enjoying it, is illusory and ephemeral. When we are not in control of change, change seems to work to our disadvantage. It is someone else's changes that seem set on putting us at a disadvantage, and wherever we look someone else always seems to be working to their own advantage.

On the world stage we can look back at the major changes that have affected our prosperity. When the threats emerged, we never saw them. When the threat was realised, for some industries, it was already too late. To survive, and be ever more successful, means leading from the front and having others follow.

Western industrial nations eclipsed by the Rising Sun

The rise of Japan since the end of the war is well documented and the results of its rise surround us all in our daily lives. Although one can argue that we will not be caught napping again, or that such a situation with another country is unlikely to occur in exactly the same way, the experience has a number of lessons that should not be easily forgotten. Despite our success at innovation and being first in the market with many products, the inhabitants of the western industrialised nations are reminded daily through the use of our video and fax machines of the dangers of complacency.

In defence of the position the West found itself in, some commentators suggested that the history, religions and culture of the East predisposed them to making the transformation in business approaches, the results of which the West now suffers from today. Examples of Japanese companies operating in the West, with similar transformed results, now provide sufficient proof that there is nothing ethnic about taking the long-term view, empowering the workforce, enabling people to take pride in their work, delighting customers, removing process variability, or about lowering costs to invest in growth.

The message is clear. If the same workforce can produce such results, then the change in management approach must be at the heart of the transformation.

Eastern Europe, the back door into western Europe

The fall of the communist bloc and the rise of the independent nations surprised many. No sooner did the Berlin Wall fall than Germany was re-united. When the Cold War ended, western defence industries suffered the rapid consequences of peace breaking out.

Attending conferences in East European countries provides a salutary lesson in not underestimating the collective eagerness to throw off the shackles of seventy years of mediocrity and a desire to learn avidly all the skills to become a free market economy. One of the roads to prosperity is seen as entering the community of west European nations, and the ticket to get there as being the accreditation to all the necessary European and world standards, and in particular, passing through the golden gate of ISO 9000.

Although we can remain complacent in the knowledge that accreditation to an internationally recognised quality standard does not guarantee low scrap rates or products that we in the West would want to buy, it does demonstrate that the East is beginning to remove what it would see as initial barriers to gaining entry to western markets. In the meantime, a low cost of living and thus manufacture costs, provides the base from which quality products will be made and then sold in the West at prices that will be keenly competitive.

The roller-coaster of feast and famine

In the heady days of the early 1980s, many economies were buoyant and the entrepreneurial spirit was in the ascendancy. Growth was encouraged with many new businesses being created. When the crash came it hit hard. Against the predictions of many pundits, and against the desire of government, the recession spiralled deeper and deeper. Businesses collapsed as confidence slipped away and spending was arrested. In the recession of the early 1990s, its depth and longevity brought some sectors of the economy face to face with an experience they had never dealt with before.

Some sectors were hit hard for the first time. Banks, building societies and insurance companies had already found that with legislation allowing them into each other's territory, the loss of a protective shield, coupled with a declining number of potential customers, left many poorly placed to respond to the rapidly changing economic circumstances.

Lessons from the world stage

History tells us that although many lessons are learnt during each economic cycle, there appears to be something in the collective psyche of the leaders of many companies that lends substance to the proposition that 'hoping for the best' is a strategy practised by many. That management is a science and the skills required are a key competence to be learnt and applied is too often ignored by those that have achieved short-term results in a business through adopting techniques that have served them well in previous global economic conditions.

The certainty in business life is that the future is not that predictable. At best, some parts of the organisation's future can be controlled and the path to the future determined. Any change that improves the business disturbs the local equilibrium in the pool of immediate competitors, and some reaction from them is predictable. But one must always be on one's guard for the emerging competitor, the one that strikes from another sector or from another nation. There are no rules, nobody will behave as the perfect gentleman, no organisation has the right to be successful, survival is not compulsory.

Customer focus

In a sellers' market an organisation can almost do as it pleases. In a monopolistic situation, costs can be recovered at will and the service is whatever the organisation wishes to provide. Where the service is a legal requirement that people have to use, and no commercial pressures exist, then 'customers' take their chances with an expectation of minimum success.

Times change. Can we remember when we felt honoured to be given a mortgage and then only after we had saved for years with one building society. Can we remember when only one type of telephone box appeared on the streets. Can we remember when government departments did not issue the public with service level agreements which allow us to measure the service provided, and obtain redress when the service falls below acceptable limits. The focus on the customer is proving to be a key driver of change for all organisations in a market-led economy.

Customer perceptions are changing

The rise of the notion of 'Quality' has found its way into many people's lives. At the simplest level, 'Quality is whatever the customer says it is'. On this basis, any contact with a company provides both an opportunity to delight the customer and differentiate your products or services. By the same token, any contact with the company by an enlightened customer runs the risk of not coming up to expectation and becoming the bad experience that customers tell their friends about.

The threat and opportunity in business today is the new dimension of the discerning customer. For many sectors, customer service has become the serious contender for the position of major differentiator, even where the obvious choice has traditionally been through the intrinsic quality of the product. People have become educated to recognise that they are customers and, as such, have become aware of their status when dealing with organisations which have remained steadfastly ignorant of their need to treat the service receiver as a customer.

Cross-sector influences

The change in focus has been dramatic and the implications far reaching. Standards of acceptable service do not appear to be sector specific. Customers have a basic set of standards that can be re-set by any organisation and transferred as a need to other organisations. The implications to any business are that just to watch competitors in your own sector is to risk missing a change in customer needs that has been created by someone outside the industry.

What was once perceived to be a clear competitive differentiator can be overtaken overnight by one series of advertisements. For example, a telecommunications company advertised a one-stop-shop for a range of services – one phone call to arrange many services. The principle of the one-stop-shop gets into the consumers' minds as being a much more convenient way of dealing with another supplier; for example, an insurance company or a bank. Once a bank offers all its services by telephone, then it has clear differentiation from its own competitors, and at lower unit cost due to a lower physical asset base.

A number of car manufacturers simultaneously launched a series of money-back or vehicle exchange guarantees for customers that found any faults with their vehicles. As well as setting a new customer service standard in the automotive sector, customers' perceptions of good supplier behaviour are re-set and the requirement for the same deal transfers to other sectors and suppliers. Initial differentiation quickly becomes the minimum acceptable level.

Government catches up and forces the pace of change

In the UK, government has been behind an inexorable drive towards forcing the pace of change in the nature and level of service that the 'customer' can expect and has a right to receive. Agency after agency is running headlong towards providing customer service against published service standards. Charters are the published mechanism to change our perception of what we have a right to expect and market testing has become a mechanism for forcing a change in the capability of the organisation that provides the service.

In the way that politics looks for simple messages to capture the imagination of the public, the implementation of such changes is not without its own pitfalls. Publishing a charter raises the public's expectation of a better service. Publishing before changing the capability of the organisation leads to customers having the ability to measure the service against standards that, initially, will not exist. This exposes the front-line employees to the venom of customers, with the employees invariably lacking the empowerment to make the capability changes necessary to satisfy customers. Charters can thus annoy both the public and employees, lower the credibility of the organisation, and create a cynical mindset that devalues the drive towards improved customer service.

Charters in themselves change nothing. Only management, with the skills and competencies to know how to improve processes, working with the employees, can make any significant improvement to an organisation's capability. Charters written without due regard to meeting the needs of customers are also likely to miss the real factors that are critical for the organisation to deliver.

The change in emphasis towards customers is fast becoming elevated to being a basic human right. With this level of focus there will be no going back. With millions of additional consumers moving into an enlarged free market economy around the globe, ignoring the relationships an organisation needs to have with its customers is to invite the collective wrath of large segments of the world's population. Where that wrath is expressed through a freedom of choice, some businesses will face failure. In the public domain, the issue may well become a key factor in the choice of government.

Cultural baggage

For many years in the West many companies remained blind to the threats from the East. Even when it became visible, many companies still ignored the threat. In the face of stark economic reality, irrational behaviour still held sway and the situation was further rationalised by drawing on the reserves of cultural baggage. Some managements would claim that people from other nations are different, by some standards inferior, that they have different religions, that their society and family values are not like ours. In any event, our technology will remain superior.

These attitudes provided an inertia that kept the notion alive that what the West believed in, the values it held and the overall society behaviours that resulted, would stem the tide of imports and pull the West through. The evidence suggests that it did not.

Behaviours driven by our attitudes and beliefs

A company, business or organisation can be viewed as a system of collective behaviour that determines the relationships with the environment, society, customers, suppliers and competitors. The system of behaviour is driven by the shared beliefs and values and, to an outsider, can be categorised as the organisation's common purpose and way of doing business – its culture.

The system of behaviour collectively provides the means of achieving an end result, and the range of end results will be limited by the collective behaviour. Some end results will only be achievable by a change in collective behaviour, and thus, by implication, the constraint on change is created by the beliefs and attitudes. As these are part of the management style and approaches to running the business, then any radical change in the business performance will be hampered or accelerated depending on the level of radical change in management's beliefs and values.

Interactions within an organisation between groups with different beliefs and values are a source of conflict. More insidious and, in the long term, more damaging are the situations where individuals feel they need to conform to standards of corporate behaviour, rather than behave in a manner that is determined by their own beliefs concerning the customer relationship. When this happens, the result is invariably frustration and low morale. For most people, the influence of management's expectations is infinitely more powerful than the influence of meeting customer expectations.

This introduces a very human dimension, and presents severe challenges to our traditional thinking on management. The challenge is to harness that latent dedication, involvement and excitement that can be delivered by staff sharing beliefs and values that align with a management aligned to meeting customers' needs.

Haunted by the past

If another approach to management can now be seen to be more successful than any that have gone before it, should we blame management for the errors in judgement that led to many of the now discredited approaches of the past?

Everyday we do the best we can with the knowledge we have. Where approaches in the past have delivered leaps in productivity they were grabbed and copied with enthusiasm. Such it was that work study and productivity schemes were based on the premise that the worker had to be encouraged through rewards to give up the time the worker built in to every task in order to have an easy time.

Experience now suggests that most improvements to productivity arose from changing elements of the process outside the workers' direct control. Improving tooling, the accuracy of drawings and data, and the availability of materials, all contributed to improving the system in which the worker toiled. Although changes to the system around the worker came about through management exercising their authority to make the change, in most cases the worker remained a commodity in the centre, available only to sell hours to the company.

The relationships between management and workforce became antagonistic and unco-operative as a result of this treatment of the laudable desire to improve the effectiveness of the business. The perceptions on both sides, once coloured by the role management believed it had the right to manage, set in concrete the expectations each had of the other's behaviour. In such an environment, the role play served only to perpetuate poor management practices. To get on in such a company one only had to learn from those above and repeat the same management approach.

Inside the atmosphere of hostility, productivity improved. The hostility, in itself, never proved sufficient cause for management to question seriously its own approach to managing the business. On the contrary, the improvements to productivity were seen as attributable to management's firm actions against the tide of the workforce's reluctance to do a fair day's work for a fair day's pay.

After all, everyone could see that the same approaches were common practice wherever one looked. Any change in approach would just have been change for change's sake.

Why is it that we still hear management calling for increases in productivity from their workforce to pay for salary increases? The implication from such a request is clear; it is the workforce that are dragging the company down. It is management's responsibility to take actions that re-position the business to obtain revenue, and management's responsibility to use their authority to change and improve the processes in which people work and thus reduce unit costs. In retrospect, we now see how the economic fate of nations can become finely balanced on the scales of knowledge and authority.

Knowledge and authority

Improving the activities in a function or cross-functional process requires knowledge. To make a change requires the authority to do so. What is the interaction between knowledge and authority as you move up the organisation structure? In Figure 1.4, we see the career path that many follow. New staff start with little knowledge of the process in which they work and they have no authority to change the process. In time, staff have the knowledge of how processes work well and why they fail, but they have little authority to make changes. With promotion comes authority but those with authority lack the detailed knowledge of the processes in which their staff work. For a multi-functional process, even the staff have little knowledge of cross-functional interactions and are usually constrained to remain ignorant of the process steps upstream and downstream of their place of work. Multi-functional processes meet at board level, but the agenda of board meetings will rarely contain issues concerning the many minor failures at functional boundaries. The role and attitude of managers is thus key in terms of aggravating or alleviating this problem. Where it exists, it is a recipe for little change.

Exhortations to improve have only served to alienate staff from management. To achieve improvements, everyone must be motivated and become involved. Obtaining improvements requires that management create the environment where staff can propose change in a structured approach that is based on the knowledge of how the people interact with the process in which they work. Knowledge then becomes the essential ingredient that motivates towards continuous improvement. For years, managers have 'managed without knowledge'. If continuous improvement is a key ingredient to ongoing prosperity, then by definition, managers must learn that they need to manage in a new way. In such an organisation, managers have a new role.

We do the best we can with the knowledge we have

As we learn more we can apply this knowledge to create a better understanding of the world about us. With knowledge, we can apply new techniques to

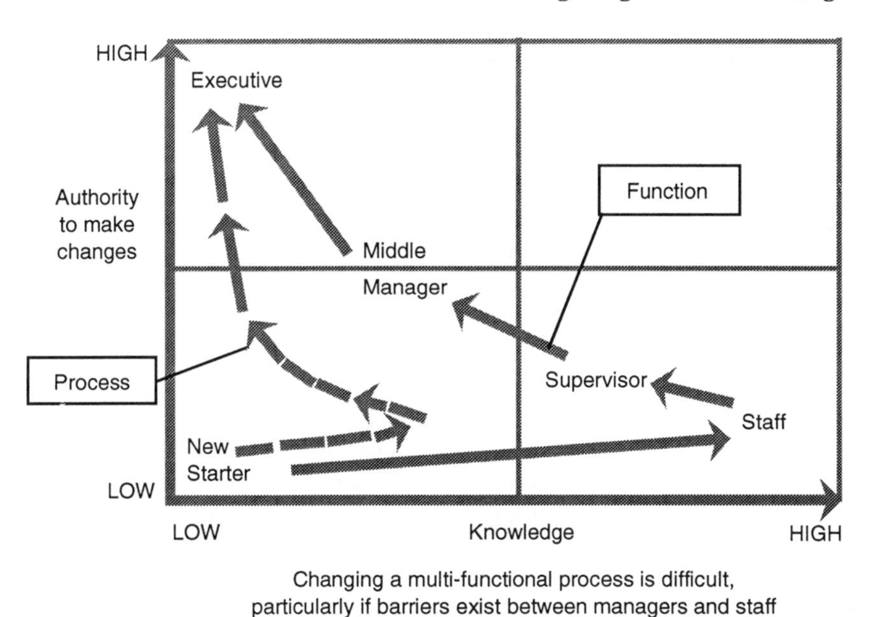

Fig. 1.4 Knowledge and authority

improve the business. If it was that easy, then every day it would be everyone's task to search for knowledge and apply it, and we would all be actively encouraged to do so. But this is not always the case. So often, new knowledge is seen as a threat. We have to accept that what we knew in the past, what had served us well, may now be discredited. This is not to say we were stupid, after all we can only do the best we can with the knowledge we have. Why is it then that the bearers of new knowledge find themselves so persecuted? It is only in recent times that Copernicus and Galileo have been vindicated and their heretical theories accepted in certain quarters.

The doctor's dilemma: a short story

In foregone days, medicine had been successfully practised without the knowledge of germs. In a pre-germ era, some patients got better, some got worse and some stayed the same; in each case some rationale could be used to explain the outcome. In any event, mistakes were buried and in general we remained none the wiser.

Today we take for granted that doctors administer to the needs of their patients according to what they learn through training and experience. They apply what they learn and believe, and interpret what they see through their understanding of the way the body works. As a professional group they do not stray too far from shared beliefs, and current knowledge keeps them on the line of accepted practice. Like doctors, we are all prisoners of our upbringing, culture, and the level of knowledge of our teachers and fellow practitioners in our own areas of skill, and thus draw comfort from sharing common beliefs. We credit ourselves as being intelligent, and in general we believe we know what we are doing.

Even 120 years ago, doctors were using common practice to advise people how to avoid catching malaria. The word comes from the French, mal-aria or bad air and was believed to be carried on the night air. Early colonists took the doctors' advice and perfumed the air on retiring to sleep. At the time, their theory of medicine had them looking in the wrong places for the wrong solutions. Other people paid for such mistakes but were sadly silent and unable to join the debate on likely other causes.

In the 1990s, managers still do the same. They look to changes in tax structure, import restrictions, government economic policy, unfair competitor activity – everywhere but to their own businesses to understand what is really happening. Like the early doctors, everything is questioned except their own theory of management.

Let us retrace our steps for a moment. Imagine that you are a young trainee doctor in the nineteenth century, hoping to become a doctor in your community. You see that doctors are successful, they are respected members of the community and visibly they have profited from the rewards of their labours. They cure some people and lose others, but on balance no one doctor is particularly better than another. In the culture of the society at the time they have the trust of the community.

Let us also imagine that during your learning period, funded by the local community, you came across the work of Pasteur and Lister. You learn of carbolic acid and its use to reduce inflammation during surgery, and you learn a new word – antiseptic. It dawns on you that these famous physicians, those you respect and have learned from, are killing their patients. Your keenness to bring this new knowledge to their attention steels your resolve to go back and tell them that because they do not wash their hands or sterilise their instruments they are sewing death into every wound. At this point you have greater knowledge, you have a new theory of germs. You resolve to go back and tell them that what they have been taught, what they are still teaching, and what they firmly believe as the accepted wisdom, is completely wrong.

Now imagine you are one of these doctors, listening to this new theory from a young upstart. You are a respected member of the community, you are successful, you follow common practice, you even pass on your accumulated wisdom at some of the most respected colleges in the land. How would you feel if your established edifice was crumbling under this apparent attack, and your patients told that you are a menace to society and unfortunate to be under your professional hands.

At the top of the business, management play the game by the accepted rules and common practice. Whatever happens there is always a plausible explanation for it, and most of the time, things are done with best intent. In 1870, doctors had a sense of social responsibility, they were sincere in their efforts to do the right things and for centuries they had taken the Oath of Hippocrates. Those early doctors did the best they could with what they knew at the time, and so do management today.

What the doctors were taught was just not good enough. Some things they did were dangerous and harmful to their patients, but they learned as everyone can, even management. The profound changes in medicine following knowledge of germs need to occur again in business with a change in managerial practice. The changes needed are not a passing fad – they are necessary for survival. If your competitors have started making the changes, they will prosper and at your expense. Where whole nations have made the changes, they are leading the world in economic performance and we ignore the risks at our peril.

The dilemma new knowledge creates is profound. Do we learn from this new knowledge or do we discard it as a threat? Like the early knowledge of antiseptic, knowledge of BPM may cause the same reaction when it is first introduced into an organisation. Are you the young upstart with new knowledge or are you part of the enshrined edifice?

MANY INITIATIVES WITHER ON THE VINE

Companies try all the latest three-letter acronyms in the hope that the next one will add the missing magic ingredient. The rhetoric of governments and the exhortations of management have singularly failed to achieve the levels of competitiveness that companies need to achieve to become, or remain, a viable force on the world stage. In the 1960s and 1970s, we were still slaves to initiatives that drove us towards ever higher levels of productivity. What did this mean in practice? Efforts focused on direct productivity, the means of production. The greatest strides were made through automation, the removal of manual activity. Much of the industrial strife during this period can be attributed to the failure to re-direct the spare capacity towards growth in home and world markets. People lost their jobs.

During the same period, the shift in the type of work in many companies resulted in a greater proportion of activity appearing in the category traditionally called the overheads. It can be argued that the categorisation of people into being either 'direct workers' or 'overhead staff' has been one of the more insidious implications of standard accountancy practice. The difference between the two groups has only ever been one of the degree of ease or difficulty to measure and properly attribute costs to products, services or customers. For no other reason, the categorisation of people led to different rates of pay, different benefits, different management attitudes. Enlightened companies have seen the insanity of this divisive approach to all employees. In reality, overheads should be seen as value-adding activities, contributing, along with the direct activities, to the overall value the customers receive.

However, the term overhead activity is still in common usage and I will have to use it with some reluctance.

As overhead activity increased so did the opportunities to introduce automation through the use of information technology (IT). In the 1960s and 1970s the phrase IT was not used. Rather, the emphasis was on the computerisation of administrative manual activity. For some companies, it remains questionable whether the savings in manual activity were ever greater than the costs of the computerisation designed to displace the overhead costs.

With this emphasis on becoming efficient, less attention was being paid to becoming effective. In other words, really understanding how the activities in every process did or did not leverage the results of the business.

Losing our way in the overhead maze

If we take the sectors of energy, manufacturing, construction, wholesale and retail distribution, transport and communication, and financial services, then between them, these industries account for some seventy per cent of all economic activity undertaken in an industrialised country. From surveys we have undertaken, the clear and consistent message was that within such businesses, much management and staff time and effort was spent correcting mistakes and doing unnecessary tasks. As a result, customer service and business performance suffered. Major improvements had to be made in the way that the corporate overhead managed resources in order to concentrate on key customer service needs and to eliminate costly wastage. The cost of the controllable overhead in most companies is a large proportion of total costs. In the manufacturing sector it represents twenty-five to thirty per cent of the total. In the service sectors, it stands even higher at around sixty-five per cent of total costs.

Measuring overhead effectiveness is not easy. Overhead staff manage, plan, administer, account, maintain, control, check, liaise, communicate, correct. It is hard to decide if overhead staff are doing the right things and that they have the tools and skills to do them right. A number of reasons exist that create this difficulty:

- Relative to direct activities, it is nowhere near as easy to measure the activity that is required to support a given level of business output.
- No accounting system is designed to attribute overhead costs in relation to the activities they support.
- It is all too easy to make any indirect activity sound crucial to the business.
- Many indirect activities, such as checking for other department's mistakes, are hidden within standard costs.
- Functional parochialism and status geared to departmental size effectively block any attempts to understand where added value actually exists.
- Without a clear understanding of the overall strategy and direction of the business, no basis exists on which to judge the relative merits of various degrees of overhead activity in different business processes.

Despite these difficulties, many companies feel that they know when overheads are 'too high'. Rather than improve the processes and re-use the spare capacity to build the business, they resort to a number of techniques that just address costs. These have a number of characteristics:

- Under the delusion that the approach is fair, all departments are asked to make the same percentage cut in costs. Re-establishing resources, found later to have been vital in the longer term, is difficult and can allow competitors into a gap.
- Setting work standards may take some slack out of the processes but often productivity improvements are made at the expense of quality. Worse still, work standards may enshrine activities that should not be undertaken at all.
- Controlling promotional aspirations through job evaluation. Gaining points for pay and position becomes a key motivator of staff and managers, and leads to the generation of unnecessary activities and layers of management in order to increase one's points score.
- Budgetary constraint and repeated squeezes on expenditure attune staff to become reluctant to pursue even those proposals which would give a healthy return.

However, the key issue is not one of cost saving but of resource re-allocation. The investment in retaining people then has a higher return and impact on profit than the savings from just reducing total costs by removing people. A better course of action would have been to:

- make savings, in terms of unit costs, but increase volumes to absorb the spare capacity;
- make savings, but utilise the resources to provide a better service through better stock availability, faster product development lead times, better customer support, improved design, and greater quality and reliability.

For too many years, the emphasis on cutting costs has left many nations poorly placed to meet the international threats in the future.

The drive to achieve Total Quality

In a survey undertaken by Develin & Partners, published under the title, 'The Effectiveness of Quality Improvement Programmes in British Businesses', it was found that many companies had responded to the threat of low competitiveness by initiating a variety of quality improvement programmes, as shown in Figure 1.5.

The awareness of Total Quality (TQ) was high but, although much importance was attached to the reasons for starting, it was disturbing to find that, despite some notable successes, the average level of improvement was reported as low, as shown in Figure 1.6. A key measure, profitability, only marginally improved. Why was this so? A number of serious concerns were reported which provided clues to the two main underlying causes.

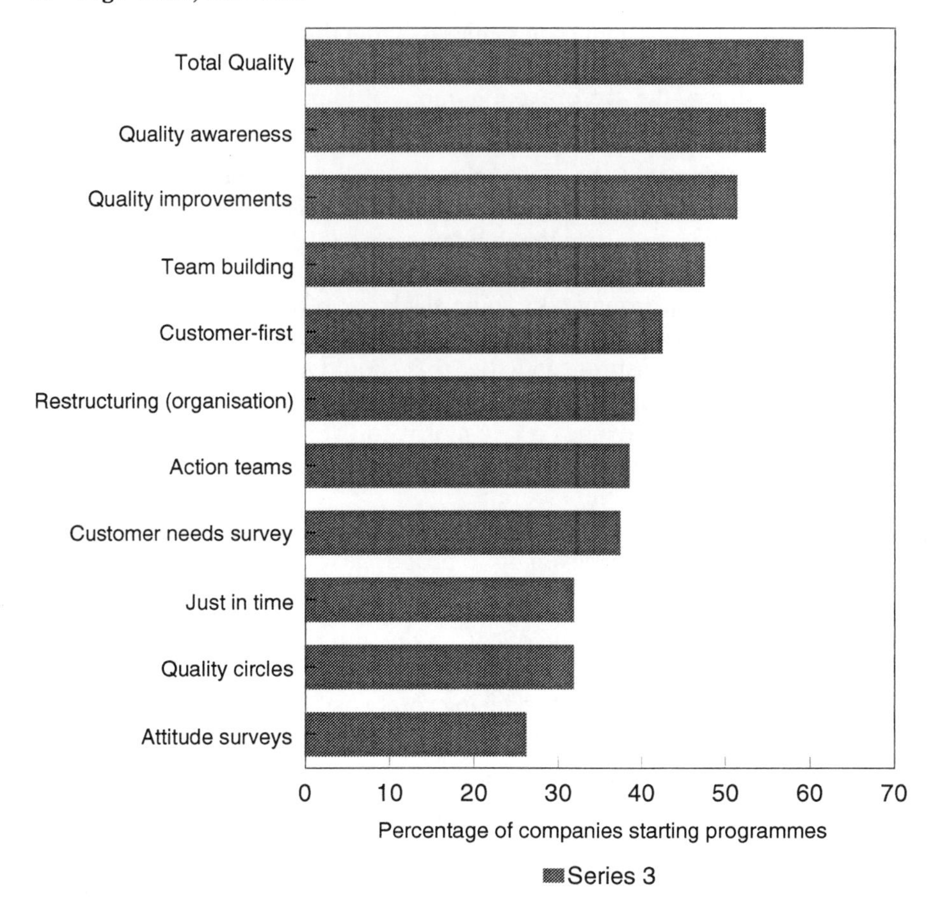

Fig. 1.5 Improvement programmes in 1990

The TQ implementation approach

The first underlying cause of reported poor success was to do with the way Total Quality was implemented. The respondents in the survey highlighted a number of factors which gave cause for concern. These were:

1. **Making Time:** Starting an improvement programme is always a problem as the effort requires additional time. Not all activities in the business add value, but the problem is knowing which activities these are and knowing how to eliminate them when everyone is already busy on everyday business.
2. **Cannot enhance internal service:** Any request to have another department enhance its service to your own, always seems to get the response that additional resources will be required. As it is difficult to eliminate wasted activity, any request for others to improve their service is often resisted if this is likely to increase costs.

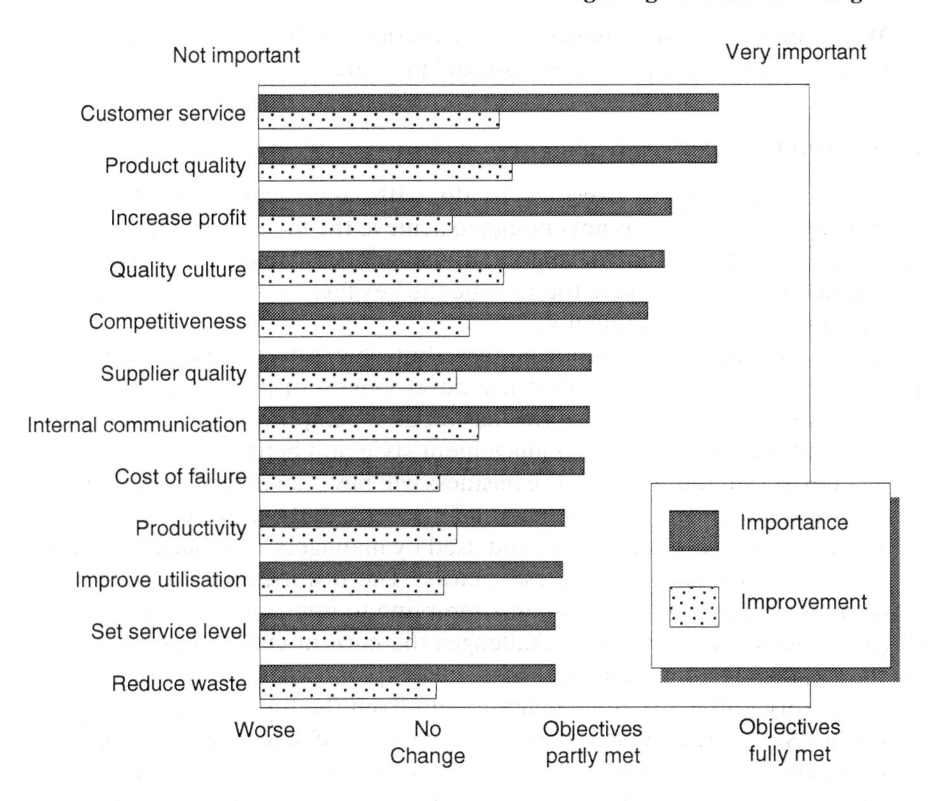

Fig. 1.6 Importance and level of improvement

3. **Few tangible benefits:** Exhortations to improve quality do not provide a clear focus on the actions needed to provide tangible benefits. When the initiative to improve quality fails to deliver benefits, then the momentum is lost.
4. **Process too long, from the chief executive to staff:** Just focusing on communicating the main message of Total Quality can take up to two years to cascade the message in a large organisation if the process just starts at the top. Only a few people receive the message at the same time, so it is difficult to track failure activities across the business. More importantly, staff know where the problems exist but the business waits too long to obtain that knowledge.
5. **Interdepartmental knowledge weak:** Staff have little knowledge of how the whole business works and thus have difficulty in tracing the root causes of problems they experience locally.

When many of these issues were experienced then the Total Quality initiative faltered and business-as-usual prevailed.

Management issues

The second underlying cause is to do with management. In fact, only addressing the first cause is not enough to achieve the motivation that ensures a permanent result. Culture change and achieving a demonstrable change in management behaviour were found to be the key factors obtaining a successful implementation of Total Quality.

Much is said and written about 'culture', but what is the culture of a business? It can be summarised as the historical development of the business since the day it started and the influence this has on all its employees. This results in a degree of commonality in the management style and behaviour where beliefs and values are shared. In total, the business will have developed a way in which knowledge is obtained, analysed and interpreted, and will have created clear limits to the authority that is held and used by managers. To change a culture, then, requires changes to all these factors. The historical inertia within any business is a major constraint when attempting to implement change, and the degree to which Total Quality challenges the current culture determines the resistance management and staff may well express when the board launch a Total Quality initiative. Often commitment from the top will be voiced but if this is not supported by action and demonstrable changes in the executives' own behaviour, the rewards and benefits of change will remain an elusive entity. A culture change by management is therefore the turning point.

THE WARNING SIGNS OF FAILURES IN CURRENT PROCESSES

If a business process is failing then what is causing the failure? We can think of errors, mistakes, misunderstandings and so on as being viruses that infect processes. Once a virus is in the system its presence is felt and if the system is not treated to become immune to its effect, then the virus escapes and infects others. A system that contains viruses is one that produces a variable output; the results are no longer predictable, stable or of high quality.

Like any illness, though, do we always know what we are treating? The effect of the virus can create serious symptoms and our response is most often determined by the higher visibility of the symptoms which divert us from attempting to treat the root cause. What is important is to be able to recognise the difference between symptoms and causes, and then, with knowledge and sufficient time, treat the root causes. What often mitigates this is a traditional measure of a good manager; an ability to make snap decisions on little evidence – the 'fire fighter'. To quote Abraham Maslow, 'If your only tool is a hammer, you will begin to see everything in terms of nails'.

All businesses contain this virus of variability. We attempt to do our best but the system consistently beats us; things just do not happen as predicted every time. Whatever we do, the process contains the virus – parts do not fit together every time on assembly, invoices have mistakes on them, specifications are incomplete, the computer breaks down, the materials are often inferior, things just keep letting us down. The processes we are using are not capable of doing the job. And, if the process is not capable, then despite our best efforts our output will be variable and, in the broadest sense, of inferior quality. Only by working on the process can it become capable and its results stable and predictable.

When processes fail our behaviours become abnormal. We run around like headless chickens, we try to drive through improvements using performance measures that confuse staff, we invest unwisely in information technology, and we overburden the organisation with accountancy practices to get a firmer grip on the numbers. We retreat into the comfort of the safe havens of our separate functions, and when all else fails we resort to exhortation, 'to do better', and 'be right first time'.

One man and his virus (another short story)

Earlier in the chapter we discovered aspects of the doctor's dilemma and the beginning of a new understanding of the effects of viruses. Let us continue the medical analogy to explore some of the issues concerning people and their interactions with processes when the virus is at work.

Meeting functional objectives creates victims

Imagine that you are the chief buyer in a business and management have set you objectives against which your performance is being measured. You are going to be measured on your ability to screw down suppliers to a low purchase price. There seems nothing wrong with this – after all, you were trained to know what to do – so you set about the task using all your knowledge. Eventually you find two cheap sources of supply for your steel bar, each supplying fifty per cent of the total requirement. You achieve and keep low prices by constantly threatening to move the balance of your orders from one supplier to the other. Both supply to specification but each supplier is at the extreme of the tolerances on material hardness – opposite extremes. To give you a low price, each supplier cannot afford to maintain its consistency close to the mid-range of specification. The material is infected with variability.

In your material stores the materials are held in the same place, as they are both to specification and thus pass materials inspection. The material then begins to infect your machine setters, they are forever changing the machining settings to allow for the different material hardness. Tool wear is now unpredictable as are the visits by the maintenance engineers. The safety stocks of tools in the tool stores have to be increased to cover the increasing unpredictable demand; the virus has now infected the stores.

The previous preventive and predictable maintenance of the machines now changes to fire fighting, and more skilled men are required to work on the now unpredictable and frequent calls to deal with unusual vibration. With more skilled men to train on a wider variety of potential problems, some begin to receive less training than others. The training department is now infected with the virus. Eventually, people with variable skills attempt repairs where they have insufficient knowledge and experience and the standard of repair deteriorates.

The virus then infects the personnel department which attempts to address a perceived problem of diversity in the ability of the maintenance engineers. Following the appraisals by the supervisors, the personnel records now include details of the below average workers. In the management accounts, tool stocks have gone up, usage has gone up, maintenance costs have gone up, but as this is a new stable level, the standard costs are adjusted and the root cause of the problem disappears from view.

You met your objective as a buyer but despite the low raw material prices, overall costs are too high and the business fails. Now redundant, you sit at home and ponder the things you believed in. You did nothing wrong, you did your best, you met the company's objectives, you followed the rules – in fact everyone did.

Surely the service sector is about customers?

Imagine now that you decided to make a fresh start, something different, nothing to do with manufacturing. A job which is simple, where you can meet people, meet customers, little bureaucracy. Ideally, somewhere where things cannot easily go wrong. When you read the job advertisement it seems ideal so you apply and get the job working in a TV rental showroom.

Things start well as the shop is always full of people and there are always at least three people waiting to see you. A lot of people come and go, maybe they are just browsing. It amuses you to note that some people just cannot seem to be able to wait patiently while others are being served. You seem to be kept busy all day so that must mean you are doing well.

All the time a new virus is in the shop just waiting. The rental forms seem a bit complicated to start with but you help the customers whenever you can. You don't have to help them when they fill in their name and address. However, if they want a TV set, insurance, credit terms, service arrangements or particular delivery dates, then they have to write their name and address six times. Customers get lots of practice writing their names so they are bound to get it right. Unfortunately, that is about all they get right.

The pressure of the queue means that you often leave customers to get on with the form filling bit and the customers don't want to bother you – you seem awfully busy. They do their best to complete the forms, drop them on your desk and then leave the shop. The completeness and correctness of the forms has been infected with the virus of variability and the process of satisfying the customer has gone wrong, right at the start. You did not know, nobody knew, that across the country a total of thirty per cent of the forms contained errors. You were doing your best, what else could you do.

Unknown to you, the virus then went on and infected the regional offices. The second from bottom copy of the six-part order set went to the regional office so the delivery of the TV could be organised and a check for stock-outs made. Many hours were spent by people dedicated to check for errors. Although some errors were found, some were not seen as errors, and when the data were keyed into the regional computer, some new errors were created. The preferred delivery date for the customer was also keyed into the computer, but as this failed to create full vehicle loads, some delivery dates were changed. Distribution was measured on the basis of vehicle utilisation so this was put ahead of meeting customer needs. In the regional office they hope that someone will be home when the TV is delivered. A mutated virus had now infected the regional office and the installation engineer and now waits to infect the customer.

The failures in the process are now exposed to the customer. The costs of these failures have to be paid for by someone – the customer. You may be lucky, all your competitors are just the same – well aren't they?

Back in the showroom you begin to notice more people seem to be coming in. They don't browse, they don't leave, they wait patiently for their turn to complain until one day your patience breaks. None of the complaints were your fault, you were just doing your best. However, you know better than to complain to your manager as you heard that is why the last person left and the job became vacant in the first place. But you have an idea: when people come to you to complain, you give them the phone number of the regional office. Better still, you put a big notice in the window, 'Complaints – ring the regional office'. Soon nobody comes into the shop to complain to you and you soon forget there was ever a problem.

In your shop things get quieter and at last you can cope. You are amused to see a competitor open a similar shop across the street. What a fool, you think, everyone round here has a TV. Shortly afterwards, your shop closes but your competitor's doesn't. Redundant, you sit at home and ponder the things you believed in. You did nothing wrong, you did your best, you met the company's objectives, you followed the rules; in fact everyone did.

You continue to ponder. While you were doing your best, management were doing everything wrong. They did not realise that their job was to manage the process. Not just the process in their own departments, but the process of the whole business. They had to recognise the viruses and how the processes were infected by them. If they had spoken to other managers they could have tracked the virus, but to do this would have required co-operation across the functional boundaries.

Everyone follows the rules

Where fear is the norm, or no two-way communication exists, or where reward schemes mitigate against team-work, then the chances of finding the faults in the process are very small. Similarly, staff are often afraid of highlighting where the process is failing as it is usually seen as a reflection of their own weakness. Some managers never invite their subordinates' opinions of the process. After all, they are managers and what could they learn from the people doing the jobs? When mistakes are common, management tell everyone to get things done right-first-time as if they think everyone came to work just to get things wrong every time.

But can you blame management for the attitudes they hold? Managers and their staff do their best with the knowledge they have. They meet the company's objectives, they follow the rules – in fact everyone does.

No knowledge creates victims

Recognising a virus, its symptoms and root causes is one thing. Making a process robust and immune to its effects is an entirely different matter. Processes are multi-functional and suffer from noise at the interfaces. The culture may actively discourage gaining a cross-functional awareness and the knowledge base may be so low that nobody is really able to spot a virus, its symptoms and root causes. In this scenario, authority to change a process resides with the executive board and few people are able to know how to improve the process.

As parochialism increases and staff have increasing difficulty in communicating across the organisation, the chances of understanding and resolving multi-functional problems reduces. More insidious and, in many ways, more alarming is the constraint on change imposed by the relationship managers may have with their staff. A poor relationship swiftly decouples the manager from the very people who have the most knowledge of the process and, particularly, its failures.

Where processes lack robustness they will be susceptible to the viruses from others. While remaining unaware of this effect, managers see only failing outputs as being failures of their own staff. This perception can also be reinforced by the department's internal customers pointing out the level of errors to their own highest level of authority. Come appraisal time, managers will have carefully tracked the errors their staff have been making – 'well, aren't the failures all their fault?'

An impasse. Management is blind to the issues, but with maximum authority to improve the process. Staff with knowledge of the issues are unable to break through the constraint of management. Staff have become victims. The variability of the process, a process the staff have no authority to improve, has created winners and losers. Appraisal is simply a lottery with your chances of losing increasing with the passage of time. Such an environment is one of fear, and although fear conjures up images of someone who is personally frightening, this scenario is worse. This fear is less tangible, it is a cultural issue and represents a behaviour pattern in the whole business that becomes the way of life for everyone. In the end, staff just give up. Why shouldn't they?

Intuition and emotion are not enough

In some companies, the measure of a good manager is the ability to make quick decisions based on the minimum evidence of a solution. Whatever problems are thrown at them you can get a snap decision back. What was the skill that was used? Was it intuition, emotion, gut feel or simply just experience? If the problems have been around for a long time, then was the experience really of any use, particularly if the problems are recurring? As shown in Figure 1.7, such a style of management avoids the very difficult bit, that is, finding out what is really the root cause of the problem. Company executives need to recognise when their managers lack the knowledge of the damage their style creates, staff without knowledge of why they have become victims, and everyone without knowledge of how to analyse the processes in the business.

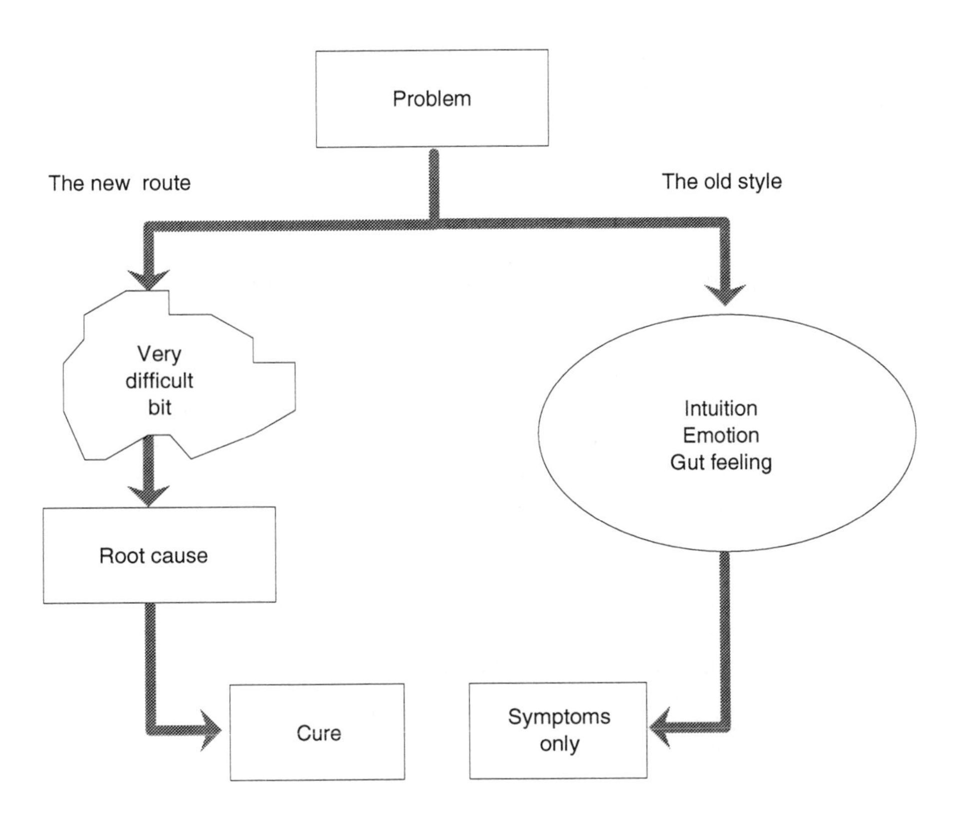

Fig. 1.7 The old style avoids the very difficult bit

Managing by results and poor knowledge

Many businesses are controlled by visible numbers; those numbers that appear monthly in the accounts. In a company practising BPM, most measures of process performance will be non-financial and the key measures will be those that impact on customer-perceived performance. A change to real measurement of the processes and their relationship to customers will be seen by staff as a meaningful method of gaining real knowledge that leads to growth and success for the business, rather than providing meaningless financial numbers that do not indicate where or how to make process improvements. Such a change in the means of measurement is a boost to motivation as it changes the focus of everyone towards the customer.

Where there is control by numbers one finds insidious practices that demoralise and degrade staff. Many people in companies are given targets as a means of motivating them to ever increasing levels of performance. On receipt of a new target what do most people think? First they make a judgement on the target's reasonableness; a one per cent increase they would scorn, a fifty per cent increase they would laugh at as impossible. When the target is around ten per cent, in a subjective sense, it is reasonable. However, if the individual knew already how to achieve the target then that level of performance would already have been delivered. Without changing the capability of the process then the process will only be capable of delivering the previous result.

Setting a higher target is just frustrating to the individual who receives it. The means to improve the process are never communicated at the same time. To ask one's manager how such a new target can be achieved is perceived as admitting likely failure, a common measure of an individual's personal performance.

To achieve a target, other than by random chance, is to encourage staff to use their ingenuity to fudge the numbers or take actions that are not in the best interests of the customers. Where arbitrary targets are at risk of not being achieved, a manager has only to link the staff member's personal appraisal to achieving the target and the result will be immediately achieved.

Targeting staff on volumes always puts staff into a dilemma. Should they meet the needs of customers or should they reach their arbitrary numerical targets? This dilemma has legal and regulatory implications in the finance sector where cross-selling of products to meet targets conflicts with the obligation to give best advice. It is also noticeable that an aware customer who discovers they had less than best advice is likely to shorten the relationship and then tell others. Over the years, this practice has been extraordinarily successful at providing a number of key results:

- quick short-term adjustments to business results;
- long-term destruction of customer relationships;
- diversion of staff ingenuity away from process improvement and away from building customer loyalty:
- undying disrespect for senior management;

- a route for senior management to abdicate from their real responsibility to improve the capability of the processes;
- a convenient mask to cover up the truly remarkable business results that would be achieved by other means (because of the quick short-term adjustments to business results);
- the destruction of the basic integrity, beliefs and values of the staff who are the victims of the practice.

Another measure that has a similar effect to those listed above is the ranking of performance of similar groups, typically, branch to branch, area to area, one police force to another, one school to another. Ranked lists have a special characteristic; someone is always at the top and someone is always at the bottom. Further, half the list will be above the overall average and half will be below. Clearly, the lower group will be seen as inferior and will remain in that category however much all the groups moves towards excellence relative to the competition.

The ranked performance approach encourages half the groups to attempt to discover how to get into the top half (at the expenses of someone else already in the top half) and encourages the top half to remain complacent to the need to improve and the need to learn continually. It is surprising how much untapped market potential really exists in the areas served by the groups in the top half of any ranked list. It is often the case that where the ranked list is used as a performance measure, the actual variables that determine performance are nearly always outside the immediate influence of the groups being measured. Other than lassooing customers off the street, a building society branch can do little to influence the volume of sales of mortgages. Macro-economics (a recession, for example), the population a branch serves, the relative location and strength of the competition, and the product's branding in relation to current customer needs are the key influences on the volume of sales.

In any event, the competition is strongest between the groups within the business and not with the external business competitors. Sharing best practice is the casualty of ranking groups, as to share best practice is potentially to let one of your peer groups get ahead of you on the ranked list.

In the absence of knowledge of a better way to grow a business, what else could managers do? It is unfortunate that in national economies that are measured by short-term results, little motivation exists to change the practice. How depressing for humanity that things ever got to this state.

Setting traps for customers (a personal experience)

While passing through one airport terminal at 7.30am one cold morning, I was lured by the smell of warm food to a new 'Breakfast Emporium'. It seemed popular; four lines of customers were slowly moving towards the counters where four smartly dressed servers were in active conversation with each customer at the front of each line.

I had plenty of time, and management had thoughtfully put up large pictures of the range of breakfasts one could choose from. From the back of the queue customers were able to start making their selection. This was both a time-saving mechanism as well as providing time to balance the opposing forces of one's rising appetite for a full cooked breakfast, encouraged by the delicious smells, against my conscience that suggested a toasted sandwich of tomatoes and mushrooms would provide a healthier meal.

As we slowly worked our way down the queue, our choice could be further amended by the prices that now came into focus. As well as health considerations, we now could balance our judgement on value for money criteria. The time in the queue was well spent, as buying a breakfast is a serious business.

On arriving at the head of the queue, all the conflicting forces had been resolved into my personal choice. An English breakfast would have been less flagrant than an American breakfast but this was further tempered by the knowledge of the forthcoming lunch appointment. I chose a filled roll of bacon and mushrooms. The perfect compromise, and I looked forward with delight to its consumption. I placed my order for the filled roll of bacon and mushrooms only to have my hopes dashed by the reply. 'Can't you read the notice. Filled rolls only after 11am, cooked breakfasts all day long!'

This posed a dilemma. Pay double for a breakfast I didn't want. Pay nothing and walk away, then join a queue somewhere else. Argue the point. In the event I ordered the cooked breakfast, sat down and watched the behaviour of new customers. The reason for the long queues was now apparent. The servers' time was spent telling customers they couldn't have what they had chosen. The people at the back of each queue all actively spent the time choosing from the large pictures and once they had made their selection, they never looked at the small notice when they got to the head of the queue.

I made a quick customer survey. A third of the customers had what they chose and their needs were met. Reluctantly, a third had something larger and more expensive. A third walked away.

I joined a queue again, the queue being served by the manager, who now had to help as the queues were getting longer. On arrival at the head of the queue I asked if the equipment was any different to make the smaller breakfasts. Did it take longer to make the smaller breakfasts? Was there any technical reason at all why only half the menu was available at breakfast time? 'No', was the answer. I made the manager aware of my findings concerning the level of disgruntled and lost customers and asked if this was head office policy. I received the standard reply of a victim. 'I don't know, I only work here.'

Complaining to victims just makes them bigger victims. Later, I wrote to their chief executive pointing out that their policy was having a net negative effect on income. No doubt the head office view of the policy was to force virtually captive customers into having a higher value breakfast at the time of greatest desire and thus increase revenue. Such stories can have happy endings if the outcome results in a change. This one doesn't, no reply was received. If pressed in private, I gladly give people the company's name.

The point of the story is to highlight two key issues:

1. That decisions made far from the customer interface are likely to lean towards managing the margin rather than understanding the value of long-term customer retention and the free advertising effect that comes from delighted customers who advocate the service or product to others.
2. That many employees get mutually exclusive signals from customers and their own management, the degree to which they serve one making them victims of the other.

Performance versus putting customers first

A supermarket found that one of the side benefits of its new checkout scanning system was the ability to measure the scanning frequency of each checkout operator. These results were tabled and found their way into the local productivity scheme; poor results leading to poor appraisals. The message from management was clear: it was more important to scan quickly than to adjust the scanning rate to match the speed customers could pack their goods. The result? Customers were forced to pack their crushed goods in a vain attempt to keep up with the high scanning rates. Slow customers were treated with impatience from the staff. The performance measures created behaviours contrary to achieving customer retention and advocacy. All too often, things are measured because they are easy to measure. Technology can often aggravate this effect.

A telephone installation company improved its customer-order to line-installation time from an average of two weeks down to two days. While watching the installation engineer put in a new line, the customer remarked on the improved service and told the engineer that another order would be placed shortly for another line, but the installation date would not be for another three months. The engineer was aghast and pleaded with the customer to delay placing the order until two days before the required installation date. The reason? The long delay between order and desired installation dates would appear as poor performance on the part of the engineer. The customer became inconvenienced in order to overcome the poor measurement method, while being delighted by the actual performance of the service.

The 'headless-chicken' syndrome

Imagine a company with a certain design capacity of machines or people. The planned throughput capacity was eighty per cent of the design capacity, the balance of twenty per cent allowing for the unexpected – unplanned breakdowns or losses, or a surprise urgent order with a short lead time. A forward order book that shows next month full to the throughput capacity would be ideal. Further in the future, the order cover would decay over some period to zero, as is shown in Figure 1.8.

Imagine that a customer placed an order now but for delivery at the point of

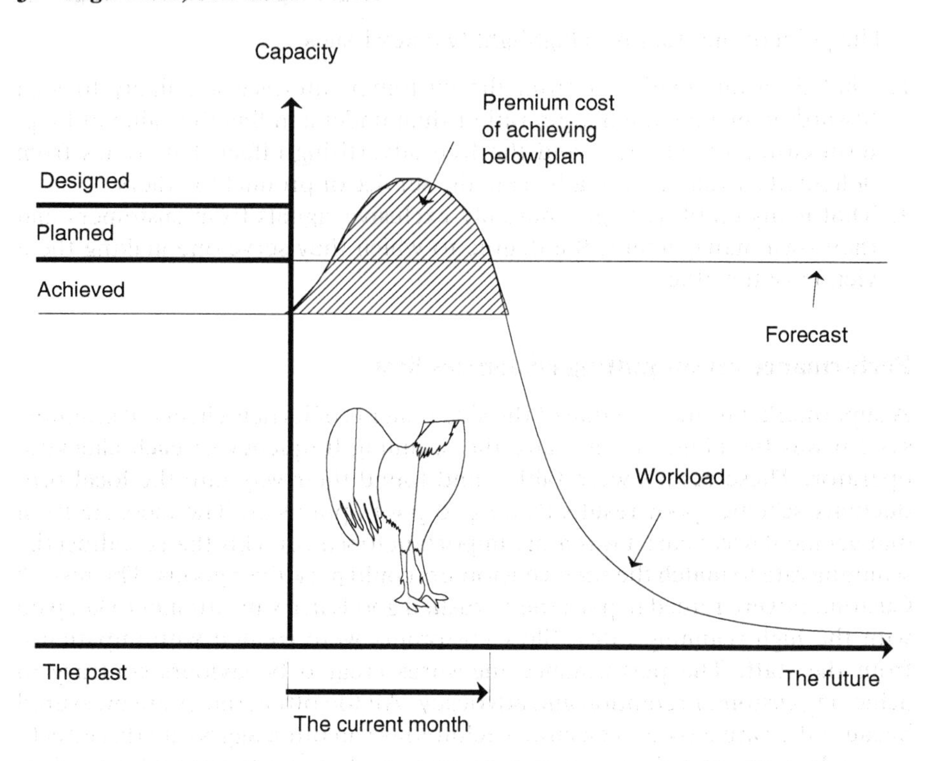

Fig. 1.8 Why can't we ever get ahead of the bow wave

zero order cover. What assurance could the company give that delivery would be on time? Clearly, one hundred per cent certainty as no other orders were in the forward load. However, as time passes and the order book fills up, this order is somewhere among many others.

In the current month or period we are likely to find that the company is in a maelstrom: things are not going right, there are panics, parts are missing, suppliers are letting them down, staff are not turning up, priorities are being changed. Everyone ends up running round like a headless-chicken. This approach has a number of effects:

1. The capacity one has to find in the current month ends up higher than the planned capacity. This shows up as additional shifts, temporary staff, additional overtime, cancelled preventive maintenance, stress and loss of patience.
2. When one looks back in time, the actual throughput is nearly always stable and often at a level below the planned capacity.

In other words, there are two losses contributing to high unit cost. Somewhere in the maelstrom of the current month will be the order placed months ago for delivery in the current month. Will it be met?

In such a business, people work hard but get nowhere. Unit costs are high and someone will be busy every month trying to reduce them by limiting expenditure. Everyone will be looking at the forward order book, grateful that in a few months' time order cover declines, but it is as if the 'bow-wave' of the current month is pushed ahead of the business and nobody can get ahead of it into the calmer waters.

The interesting thing that one can conclude is that a firm promise can be given to customers, and that is: 'Over any future delivery horizon, and particularly when we have little order cover, on the day, we will probably fail to deliver on time!'

Putting all your trust in an IT solution

A dry goods hot beverage company, mainly preparing and packing the products, implemented a computer system that printed the customer invoice on receipt of the customer order. The justification for the system spoke of 'impact on cash flow'. Measuring the transactions in the accounts department indicated that half the paper going to customers was credit notes to account for split deliveries and substitutes.

The root cause was tracked to the product managers in marketing. Their volume forecasts were linked to their promotion budgets. They received more promotion budget if forecast volumes were high, which resulted in over-optimistic forecasts to obtain more product promotion budget. The budget was fiercely fought over within marketing. Production knew this was happening, so always planned throughput based on their knowledge of the market, mainly based on what they had made in the past. Scheduling and purchasing also knew better than production and re-planned the delivery of parts and raw materials.

All this resulted in the business making what it could on the day, based on what was available to use. Each day's production went to the finished goods warehouse. Attempts were made to pick the actual orders from the finished goods warehouse, with a resulting poor match to the already invoiced customer orders. Although the quality of the products was high, the customers also scored delivery success, in terms of complete orders, and took account of the number of errors from suppliers that infected their own processes.

Isolated improvements to capability, cost justified in isolation, rarely improve the overall business effectiveness if the whole chain of dependent processes is poorly understood. This company now sees forecast accuracy and pipeline management as the key to applying technology to achieve competitive success.

Cross-charging, profit centres, cost centre budgets and minutiae

For centuries, standard accounting practices have served businesses well so we can feel comfortable that the businesses are behaving legally, that they can report their results to shareholders as a fair reflection of the overall business performance, that they pay their employees accurately and on time, that they collect what is due from their customers and pay what is due to their suppliers. This is all we ever wanted from the finance function. To do all this imposed few burdens on other functional managers.

However, other managers wanted to know how to improve processes by measuring the variation in processes. They wanted to know how and why they were keeping or losing customers. They wanted to know how often customers were telling others how good they were. They wanted to know if their staff had all the necessary competences to do the tasks required of them. They wanted to know where best practice was and how to apply it in other areas. They wanted to work together to understand how processes interlinked and the cause and effect of viruses that got into them. The needs of managers for this type of information have always been perfectly reasonable. Why is it then that so few managers ever have the time to get the information they really need to improve the business?

Maybe they are always too busy massaging budgets, trying to explain output variances, trying to unravel why certain cross-charges have been placed on them by others, trying to pass off costs by cross-charging others, trying to be measured as successful by being seen as outperforming their peers by keeping knowledge of best practice to themselves in their own cost or profit centre. Inside businesses, a beast seems to lurk that gets a life of its own feeding voraciously on information. The more useless the information is, the hungrier it gets for more. The more powerful the beast becomes, the more we become slaves to its demands and the more fearful we become of expressing any desire to question the wisdom of its never ending appetite. Perhaps the time has come to question accepted wisdom and look again at what information is really required to improve business performance and stop all other information needs that drive behaviour contrary to growing the business and achieving lowest unit cost.

Profit centres

In a research establishment, previously soley funded by the government, fifteen per cent of its work had to come from private industry – a move towards privatisation. This new 'commercial' impact drove everyone to think 'commercially'. The scientists quoted their customers for potential research projects and, on occasion, had to obtain the cost to manufacture test rigs from engineering services. The latter department was set its own objective to 'break-even'. The commercially minded management in engineering services decided to be more than a cost centre – they became a profit centre. By recovering more than their costs, they made a 'profit'. The price on the internal quotes rose. The scientists, also mindful of the new commercial environment, began to find the engineering services' prices too high and obtained competitive quotes from outside manufacturers. As these prices were invariably lower, more work was placed outside and the volume of work remaining on-site slowly decreased.

Fixed costs and lower volumes in engineering services spurred them to raise their rates to remain in profit, resulting in further declines in work. From the research establishment's perspective, total costs rose as real money went outside and the fixed costs of engineering services remained the same.

A bank established a centralised unit to provide branches with foreign currency and travellers' cheques. In line with the bank's customer-care programme, the unit established methods and processes to provide a responsive and proactive service to branches (anticipating demand at branch level during the holiday season, managing distribution economically using the bank's bullion runs, and so forth). Inexplicably, demand from the branches fell, and it transpired that branches were buying currency and cheques from competitors! The reason was simple: every branch and central unit in the bank was treated as a profit centre and was required to make a 'contribution'. The treasury function charged the unit for the currency it 'bought', plus a margin; and the transport function charged the unit for distribution, plus a margin. The unit likewise charged branches, plus a margin. Unsurprisingly, the branches found competitors' prices cheaper, and bought from them in order to keep their own costs down, even though they took longer to supply. The accounts department spent a lot of time managing all the cross-charges and trying unsuccessfully to reconcile departmental contributions with overall profit.

Cross-charging

A computer centre manager religiously set out to be cost neutral by charging out computer usage. As is often common practice, this was based on computer process unit (CPU) time linked to user departments and project codes. The cross-charging absorbed one person full time and much of the time was spent arguing with users on miss-bookings. The accounts were nine months behind.

Among the users another behaviour was observed. When people arrived at work, and before taking their coats and gloves off, they would sign-on to their terminal with their own code and first project number. No new numbers were entered by anyone during the day, whatever the work they were doing. To sign-off was to lose the line as insufficient lines existed. All postings based on codes and CPU time were erroneous. When it came to upgrade the capacity of the computer, only forecast usage based on the evolving and changing demand formed the basis for the decision, never past usage by department and project.

Cost centre budgets

A company introduced a simple method to improve the cost effectiveness of the use of taxis. By arranging to use a single supplier, any travellers only had to print and sign their name against the meter cost recorded on the taxi's log sheet. To prevent fraud, the passenger retained a tear-off copy slip of the cost. On a monthly basis, the taxi company presented its itemised bill which a clerk in accounts dutifully used to look up where each passenger worked, then looked up and noted the cost centre manager's name and cost centre number. The annotated invoice then started its journey around the site for each manager to authorise and add the cost centre allocation number for taxi journeys to the invoice. Later, these costs would be accumulated and set against each cost centre's monthly taxi budget. No doubt later, each cost centre manager would be asked to spend their time tracking monthly variances of actual against budget.

Three months later, the irate taxi company would begin to demand payment. The problem for the accounts department was knowing where the invoice was in its journey around the site. Regularly, the taxi company would threaten court action for payment. Accounts payable would just ask for a faxed copy of the invoice and pay the same day on receipt of the fax. The monthly sums were only around twenty times greater than the average taxi journey.

In a year, forty per cent of the invoices never found their way back to the finance function. Those that did return were duly posted, often twice if the faxed copy had found its way into the system. The small amounts became the source of many queries from the cost centre managers who could not reconcile their variances. Finally, someone did question whether the whole process added any value, or just unnecessary cost.

Both companies finally agreed on an annual figure and twelve self-billing monthly invoices. A single cost centre captured the posting to the accounts. Both companies saved high levels of activity that recovered any error in estimating the annual usage. A win-win relationship.

Death through innovation

Innovation is exciting, particularly product design and launch. It may be the only thing that catches the eye of the managing director followed by regular visits to the troops at grass roots level. This alone gives the innovation team a feeling of importance.

In one company, the head of New Products listed over ninety failures of other departments to respond quickly to the demands placed on them for pre-production runs, laboratory tests, customer panel sessions, and so on. These grievances caught the ear of the managing director and everyone quickly saw that to respond quickly was now an essential reaction. After all, the future lay in innovation.

Innovation and product development were the revenue for the future. However, the costs had to be funded out of current revenue. Whenever the New Product manager's demands were met it was at the expense of delivering the current products and services, resulting in squeezed margins. The arrogance of the New Products department put them above responding to requests from the other departments for an estimate of the resources they would need to provide. As a result, nobody had sufficient resources to provide services to current customers and to the New Products department. Instead of retaining competitive edge and revenue today, the company went into slow decline, eventually risking death through the impact of innovation that was designed to secure its future. Functional primadonnas, particularly concerning innovation, can presage death.

The annual operating plan, but no company strategy

Few days can be more disheartening than watching directors or functional general managers presenting their functional operating plans in the absence of a clear strategic direction to provide a framework for the future. At such sessions, no cognisance is taken of each others' plans, each person intent on grabbing the initiative and the available resources. Figure 1.9 shows the stylised responses to the chief executive's annual request to improve annual profits. Later in the year, when cross-functional issues arise and multi-functional projects need resourcing, pained expressions arise as managers find they are isolated inside their own initiatives. Staff soon discover that they have little strategic context to guide them and find they are faced with conflicting interests when attempting to deliver or request a service across functional boundaries. Quite often, the corporate projects office finds itself reduced to chasing each functional project leader just so it is able to plot progress (or otherwise) on a spreadsheet. This is a waste of the competencies of the people one usually finds in such departments.

Managing Director: '*We must increase our profits next year. What are your plans?*'

Function	*Response*
• Marketing	• New markets
• Sales	• More volume
• Product development	• More product lines
• Purchasing	• Lower purchase price
• Manufacturing	• Improve productivity
• Quality control	• Closer control of quality
• Distribution	• Better customer service
• Finance	• More analysis
• Personnel	• Re-organisation
• Planning	• Tighter control of these plans

Fig. 1.9 Another year of functional operating plans

The annual conference

Charismatic leaders who use knowledge effectively can remove mountains of corporate inertia and set an enviable example of the correct behaviours expected from everyone. Charismatic leaders, who lack a well-thought through plan for the future, damage the business as they give the impression that the future is secure while making little contribution to achieving future success.

The annual conference provides a vehicle for this latter type of leader to impress through hype and exhortation rather than substance. For a while, employees watching these performances can have their spirits uplifted. However, cynicism swiftly follows if no concrete evidence of the plan for the way forward appears. Such businesses are characterised by having vision statements that are only founded on hype.

So far, in the book, some of the reasons for starting on the BPM journey have been highlighted. Now is the time to ask yourself:

- Are some of the scenarios alarmingly similar to those you find in your organisation?
- Are you a victim of some of the scenarios?
- Are you responsible for perpetuating these scenarios?

If the answer to all these questions is 'No!', then the book will prove only interesting reading. If the answers to any of these questions is 'Yes!', then the book is necessary reading. This is not a cause for depression, but rather an opportunity to begin the process of change within your own company.

2

POSITIONING AND CAPABILITY

If you don't know where you are going, any road will take you there.
If you don't know where you are, a map will not help you.
Ancient proverb

THE INEXTRICABLE LINK

The BPM journey is founded on the principle that there is an inextricable link between positioning and capability. It is useful to remind ourselves of the definition of the two key axes that help us plot the journey, and how they are linked, as shown on shown in Figure 2.1.

Positioning

Positioning is to do with external factors such as:

- understanding customer needs;
- understanding competitor initiatives;
- determining the business's financial needs;
- meeting changing legislation;
- environmental constraints.

Positioning leads to higher levels of revenue through increasing market share, increasing the size of the market and by retaining the first-time customers acquired by the business.

Capability

Capability is to do with internal factors such as:

- key business processes;
- procedures and systems;
- competences, skills, education and training;
- attitudes, style and behaviours.

The capability is changed to deliver the positioning. Capability creates the costs

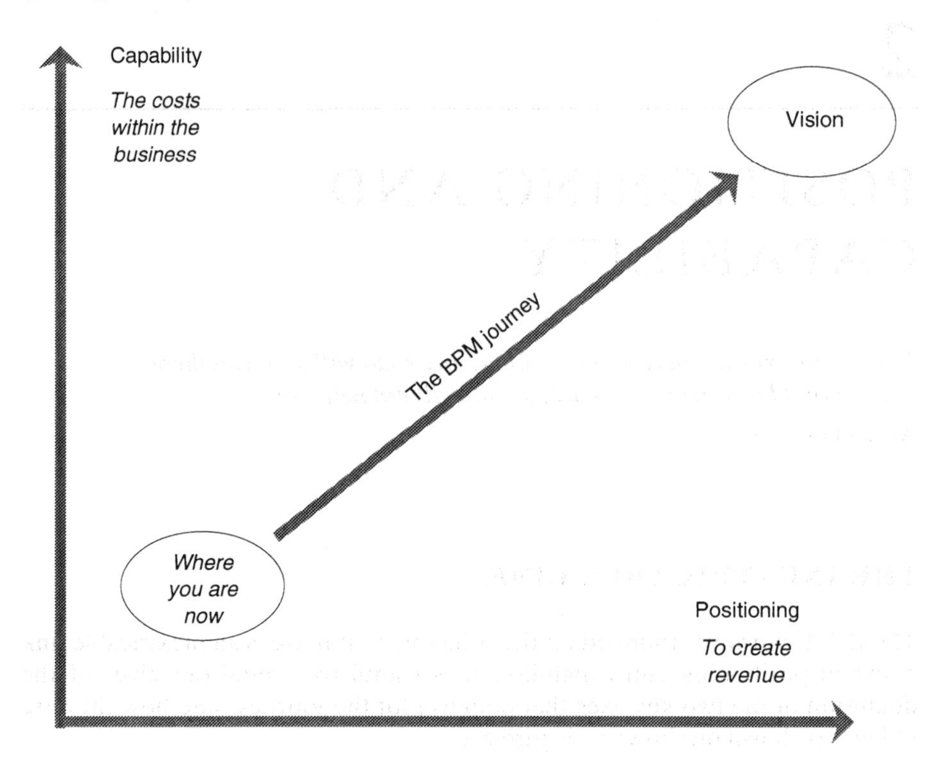

Fig. 2.1 Linking positioning and capability to maximise profit

in the business. The inextricable link joins revenue and cost. Getting the balance right enhances profits.

The Vision

The vision has to be a clear, competitive differentiator that maximises revenue and obtains lowest unit cost. The key is to have a perceived differentiation of being better than the competition in whatever terms the customers choose to measure, and to deliver this at lowest unit cost. It would be foolhardy to position the organisation to meet, or even create, a market need it could not fulfil without an unprofitable level of resources. Also, an enhanced capability to provide a cost-effective service that no customer wants is to have allocated resources with no hope of return.

Knowing the appropriate positioning and appropriate capability are the keys to ongoing success.

> *Two men were walking through the icy wastes when suddenly an enormous polar bear appeared over a hill. One man said to the other 'What are we going to do?'. The other man knelt down, took out a pair of trainers from his knapsack and started to put them on. 'That's not a solution,' said the first man,*

'polar bears can run faster than both of us'. 'But you miss the point,' said the second man, 'I only need to run faster than you!'

The final position and final capability deliver the vision for the business and the journey then becomes Business Process Management – the management activity that addresses both axes. Given that your competitors will not stand still as you stay ahead or overtake them, there is an ongoing need to review the future position and the implications on capability. The management activity that addresses both axes thus becomes the number one key business process within the organisation. It is therefore essential that a company pools together the knowledge that resides in all the functions to understand fully the positioning dimension. Reflecting this on to the necessary capability to put in place suggests that any re-engineering should not remain the preserve of individual functional heads. BPM has to start with a need to think non-functionally – a requirement that has always existed at the top of organisations but has often been an elusive set of behaviours in many.

ISSUES CONCERNING POSITIONING AND CAPABILITY

Like many initiatives, such as Total Quality Management (TQM), Activity Based Costing (ABC) and others, Business Process Re-engineering (BPR) has come to mean something in a generic sense. From our experience, BPR in practice has developed a focus on changing capability in the short term to address current issues. In contrast, we contend that the need to develop a true vision, rather than a motherhood statement hanging in every manager's office, is the job of senior management acting as a team. Using the term BPM (Business Process Management) thus encompasses a larger task – the management of the inextricable link.

Missions and visions

For a while, writing mission and vision statements were in vogue. From the many we have seen, they have often fallen into two categories:

1. Those that highlight agonising failures in internal capability. An example from one reception area I waited in being: 'We will reduce the level of defective products reaching customers from 8 per cent to 3 per cent within 3 years'. Up to that point I had a high regard for the company's products, but from then on my uncertainty increased.
2. Those that proclaim they will be ahead of everyone in every circumstance, typically: 'To be the best provider of . . .'. What does 'best' mean? By whose criteria? Which aspects are to be measured? What will it do for every customer? When will it be achieved?

Both types of statement leave staff at a loss to know what to do next and customers remain bemused. Exhortation without knowledge or without a carefully plotted path to a specified point just raises the level of cynicism among staff and managers.

In BPM, we use the word 'vision' to mean something tangible, the result of understanding positioning and working this through to changes in capability. The vision thus links the two. To be meaningful to staff, the vision has to relate to them and implies a degree of deployment and subsequent involvement that is far beyond issuing everyone with a one-line mission or vision statement on a pocket-sized piece of plastic. To be meaningful in a business context, the vision has to relate to customers. 'Being the best' can be a meaningless statement and sets the business up on an unattainable pillar, ready to be knocked down by any single customer or the media.

Capability and segmentation

If a company looks at the mix of customers and products/services it will find that a number of processes will be dedicated to one customer group associated with one product type. Other processes will have overlaps where a number of customer groups are associated with one product, or a number of products with one customer group. The link between products, customers and internal processes can be shown in a matrix, such as that in Figure 2.2.

The shaded area represents 'common capability'. This form of segmentation

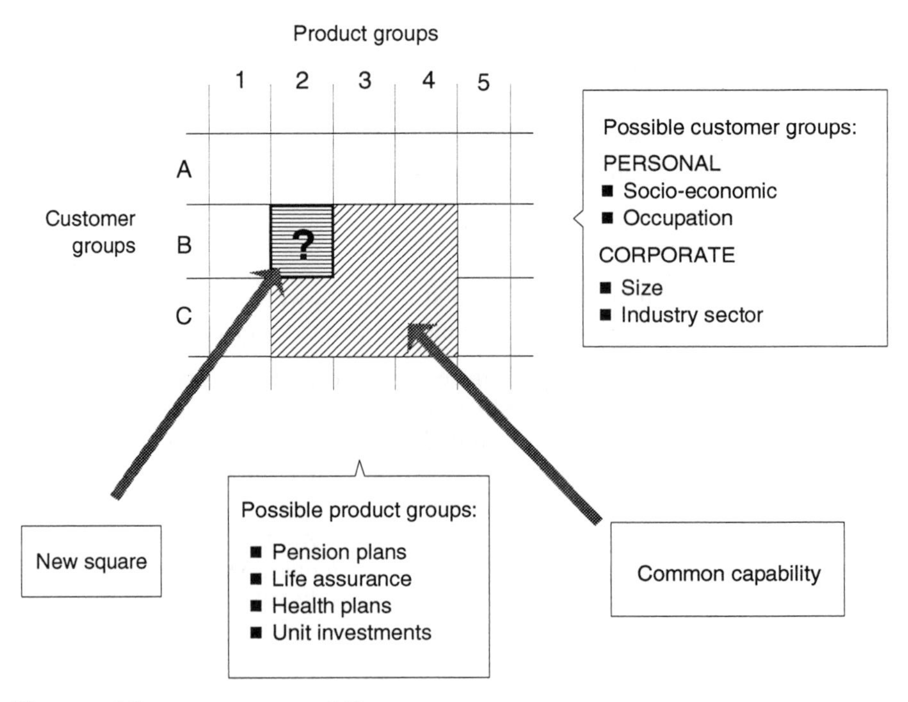

Fig. 2.2 The common capability matrix

links the internal business processes to products and customers in a way that helps to expose where interactions can reduce business effectiveness or subtly change the company's positioning. The problem comes when additional product or customer groups are introduced to the business. On the matrix, should the new square be shaded the same as the others, or should it get a different shading to represent that it does not share a common capability with the existing business? Some examples will illustrate the issue.

- A bus manufacturer made an opportunistic sale of dust cart cabs which it proceeded to build on the bus assembly lines. The characteristics of the cab, however, were entirely different from buses. The cabs were all complex hydraulic assemblies, most of which were bought-out components, the sub-assemblies fitting into large fibre-glass moulded cabs. Conventional buses were aluminium frame and aluminium sheet fabrications, most of which were built by being progressively fitted on a slowly advancing assembly line. Dust cart cabs were ordered in ones and twos over long time intervals, and whenever a cab was produced overall line efficiencies tumbled.

  The product and customer groups fell outside of the common capability that had been established over many years. The difference in build capability meant that mixing the two reduced the line efficiency of both. The true disturbance cost brought out the decision to sub-contract the whole build of the dust cart cabs.
- Automatic teller machines (ATMs) are the result of understanding changes in positioning for banks, and the need to try to delight customers by giving them access to their primary and frequent need to withdraw cash. ATMs benefit from evolving information technology and in the fifteen seconds while your account is being checked, the screen can passively advertise other services.

  In the business development function, someone saw the opportunity to take advantage of the technology and potentially captured customers outside of normal working hours. The keypad and screen could be used to sell other services by having the customer interact in a question-and-answer session in order to apply for other products and services, such as a loan or opening other types of account. The queue for the primary use of the ATM (cash withdrawal) grew, thus annoying everyone who was standing behind someone using the extended capability. Such a person would be a new customer type requiring a different product type.

  The common capability was technologically common, but constraining its use inside the original customer/product segment served only to lose that customer group. Efficient use of the assets, as seen from the bank's perspective, lost revenue. In this case, using similar technology, but streaming its use (two different types of ATM) captured a new segment without losing the other.
- A number of service providers that seem to require that queues should form, such as in a Post Office or a bank, lead customers, following logic but against

experience, to tend to join the shortest queue. The staff need to have a very broad range of capability to deal with all customer and product types. From an individual customer's perspective, there is no way of knowing the requirements of the person in front of you. Our frustration increases when we see other queues being serviced quickly and we blame the business for any delay, rather than customers in front of us. A simple change of capability, the single snaking queue, changed customers' perception of the service although the total amount of staff time to deal with everyone did not change.

This is a case of involving the customer in the overall delivery process, the queue, so that the customers' fickle interaction detracted from the very wide capability that the service provider had put in place. An improvement in positioning, more happy people using a wider range of services, came from changing the customers' involvement in the delivery process.

- For one telecommunications company, pagers were the main revenue earner. It was a simple improvement to service to connect a message service to the pager. However, the ideal package was to sell a mobile phone as well, to provide customers with the ability to answer messages. The total service had been built up over a number of years, each new service being set up as a new division. The ordering and delivery of the phone was so fraught with problems that the salesforce began to refuse to sell the total package. To further aggravate the situation, each division separately invoiced customers. In some cases, customers would pay a single amount to the parent company and let them sort it out.

  The company had assumed common capability by considering only the common sales interface and had forgotten the interactions of the delivery and invoicing processes. From the customers' perspective, they found themselves the only common link between three separate common capabilities.

In a sense, companies have companies within them, with many shared and overlapping internal processes. For some, many of the processes may require a high degree of customer interaction. We get used to conventional segmentation being something only to do with customers. The situation is more complex, and any start on the BPM journey needs to recognise the layers of complexity that can exist.

Competitive stance

So far, positioning has implied some overarching theme that will provide clear competitive differentiation and a rush of customers to your business. However, at any particular point in time, the current reality may indicate that certain customer groups are not those that you would wish to take with you on the BPM journey. Some you would wish to take with you are currently being poached by competitors. Others in the same category are being inadvertently driven to competitors by failures in your own processes. Some of your products and services may be well down their life cycle, while others are waiting to be born. Within this complexity, we believe there are a number of clear competitive stances that will co-exist within the overall future positioning theme. These are shown in Figure 2.3. Short-term positioning decisions can therefore be segmented and scarce resources allocated and prioritised to deal with the issues.

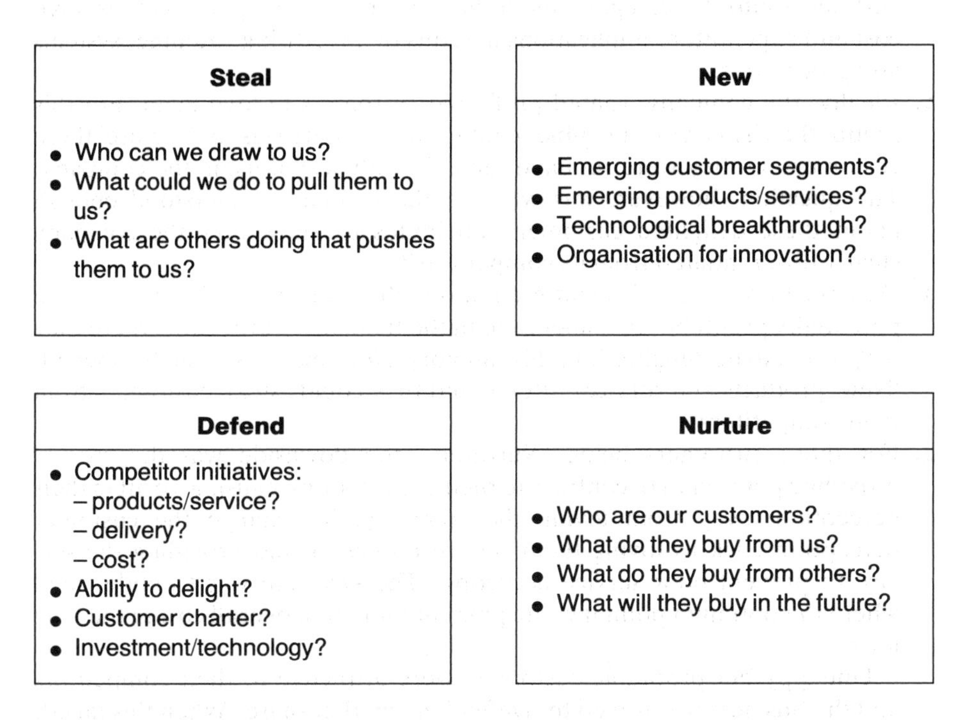

Fig. 2.3 Segmented competitive stance

An example of a need to understand competitive stances emerged in the finance sector caused by imprudent lending in the late 1980s followed by a long and deep recession in the early 1990s. The finance sector was under pressure from many companies competing fiercely to retain previous levels of profitability and wallet share from evermore discerning customers.The large swings in macro-economics, from the feast of the late 1980s to the famine of the early 1990s, had contrived to bring together declining sales with the rising costs of dealing with unprecedented levels of bad debt. Inside this minefield of uncertainty, management and staff, and the systems and procedures they worked with, strained to contain the prospect of worsening cost–income ratios. Also, the need to differentiate the business through enhanced levels of service had created competing demands for the available scarce resources needed to turn the tide in the business's fortunes. The key was to know who was in each category and why, and to know how and why customers moved from one category to another.

In contemplating how best to allocate scarce resources, a number of risks and benefits had to be considered within each customer category:

1. The overall aim was to maximise the number which were in the 'Nurture' category, the place where the company's vision had drawn them. To move customers into this category and to prevent them moving out had resource cost and expenditure implications in terms of advertising, training, systems and procedures.
2. Ideally, the company wanted profitable customers to love them. For this group, the vision was a tangible reality. The key here was to 'Nurture' these customers so they were retained, and also advocated using the company. The question, however, was whether the company understood enough about these delighted customers' attributes to know how the company clearly differentiated from its competitors?
3. With the knowledge of the differentiators, the company could appeal to the potentially profitable customers and bathe them in a warm glow where they will expect to be delighted. In this category 'New' markets could be opened, 'New' products and services offered and they could 'Steal' customers from their competitors.
4. For those customers being 'Nurtured', the downside was the risk of providing poor service, contrary to their expectations, causing a drift in their perception away from loving the company. To change the customer perception back to loving them cost resources, as did providing the service to prevent poor service occurring. The key here was to understand where the optimum point lay, the point of minimum overall cost in the long term.

   Unhappy but profitable customers were attractive to their competitors and the business was forced to 'Defend' its market share. When this failed, they incurred the costs and lost the customers.
5. Unhappy ex-customers were the most difficult to influence and worse still, these ex-customers were able to leverage the opinion of other potentially profitable customers. The size of the lost opportunity was always difficult to measure but, clearly, this category was one to be avoided.
6. Unprofitable but happy customers posed a dilemma. Often this category arose, for example, when the company issued a high-interest savings product to generate a quick cash inflow and the customers then stayed with them. Another example was a customer with a low savings balance with the habit of frequent transactions in and out of the account. The key here was to convert them to happy profitable customers or happy ex-customers, a tactic which was always risky and possibly contrary to any statements such as 'advice you can trust'.

   Unprofitable customers who were becoming unhappy formed the large group who were casualties of the recession; the customers in arrears. The key question here was whether it was worth the effort to save them or continue just to pursue the debt. Some customers were helped by easing them out of home ownership before their situation became irrecoverable. Handled well, such customers came back later as profitable customers.

7. For those customers falling ever further into arrears, they finally fell into two groups: the victims and the villains.

   The villains never intended to pay, recession or not. While the case load of arrears problems was large, the villains benefited from being less visible within the overall delays in the credit control procedures. Too many such cases, however, gave a poor impression to profitable customers who quickly realised they were paying for the villains. From the villains the company learnt to improve the credit-vetting procedures.

   The victims are those that could have been helped. Leaving aside the issue of whether it was prudent, in retrospect, to have sold them the products in the first place, the treatment of their current predicament had a key bearing on costs, both visible (chasing the debt or re-possessing the property) or invisible (the negative effect of unhappy ex-customers, or articles in the press).

   Many similar companies had been caught out by the surge in arrears cases and the common capability across the sector had produced procedures, attitudes and training geared to dealing with villains. A casual glance at many standard arrears letters showed that the collections process presumed that all arrears customers were singlemindedly intent on fraud.

   The key question here was whether the vision designed for the 'Nurture' category could be extended profitably to the victims?

An enormous leverage on profitability became the reward for really understanding the nature of the customers in each category and the costs and benefits of moving from one category to another. The essential step was to understand in detail the nature of the cross-functional processes within the whole organisation, and the impact of one department on another in the chain that eventually led to any of the customer categories.

The 'warm glow'

The positioning work may develop an overall set of differentiating messages that are communicated effectively to customers. The capability at any point of customer contact can be expressed as:

- business processes that interact and touch customers; and
- customer contact by staff holding a range of competences and a range of their own values and beliefs that translate into behaviours.

To capture these aspects we have coined the phrase, creating a 'warm glow'. Every aspect of the business needs to contribute to this warm glow and around every corner there lurks the chill wind to snuff out the flame. In Figure 2.4 the contributory elements have been divided, left and right, into those that directly interface with customers and those that have a more indirect relationship.

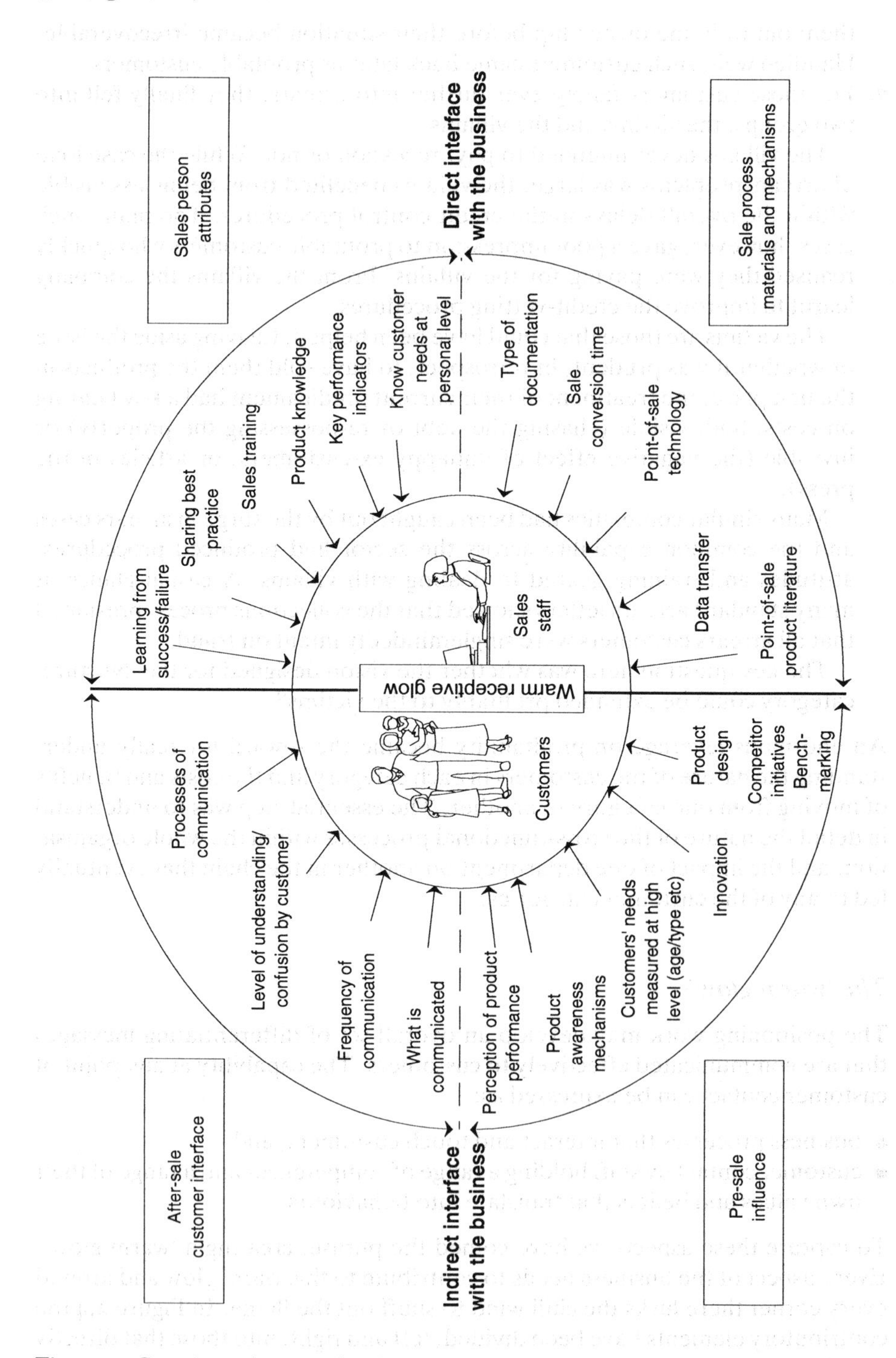

Fig. 2.4 Creating a 'warm glow'

In a vacuum of knowledge at the customer interface, we would expect to find:

- customers influenced by competitors, but that knowledge remaining unknown to the sales staff;
- sales staff influenced by their internal measures, controls and company objectives, but that knowledge remaining unknown to customers.

This situation, though over emphasised, is not uncommon. Companies often fail to see that the whole activity of getting the positioning and capability right is vulnerable at the point where expectation meets reality. Selling or providing a service is seen, in some countries, as being a menial task, or that the proposition or product will be sold despite the vagaries of sales staff.

The conflict between sales staff (and others in direct customer contact), driven by company objectives, and the needs and expectations of customers, is one of the fundamental rocks on which many company initiatives have foundered. Most of us have experienced this dilemma and, culturally, in the UK we let such instances pass too lightly. Unfortunately, staff are victims of the problem. Customers who admonish them make their lives more miserable than before. Some examples will illustrate the point.

- A colleague went to purchase a new dishwashing machine. Having found one that suited both the gap in the kitchen and the gap it would make in his wallet, he proceeded through the purchasing transaction. It quickly transpired that to avoid paying an extra amount for insurance, which was already covered by his home contents insurance, he had to select the box on the form to tick to prevent insurance being added by default. (It later transpired that a significant proportion of profits was due to commission gained from selling insurance and staff were targeted to get over eighty per cent cross-sells.) Close scrutiny of the form did not reveal any further traps. As he was about to sign the credit card slip he noticed that the total was higher than the displayed price on the product in the showroom. Patiently, the assistant explained that this was the charge for delivery.

  The altercation that followed resulted in my colleague tearing up the credit card slip and walking towards the door. Halfway there, he was grabbed by the manager who proceeded to explain company policy concerning delivery charges. The manager knew that all the competitors in the area gave free delivery but he had already lost the argument with head office. The manager offered to cut the charge by fifty per cent then seventy-five per cent. The issue was not the actual cost, which was trivial, but the whole relationship with the customer, which was designed to break down. No discretion was allowed at the customer interface, and any opportunity to delight the customer had been removed.
- Another colleague had to replace the washing machine. A house full of young children can survive only a few days without this essential item. In the shop, he was informed of the two delivery days, and in a busy life, only the

following Saturday could be accepted. At the end of a desperate week, the machine was delivered, unloaded and left, and the delivery vehicle sped off. Running the test programme, advised by the manufacturer, exposed a number of faults. The key one being that it did not wash clothes.

The telephone conversation with the shop manager produced the interesting knowledge that the drivers all finished at 4pm and nothing could be done until the following Saturday. The altercation that followed resulted in a driver appearing with a replacement at 11pm that same evening, after the driver had been out for the evening's usual entertainment. In a good mood, the driver installed the machine and ran the test programme. Everything worked perfectly. 'But why do you not run the test programme for each delivery", my colleague enquired. 'Because, we have tight delivery schedules and we have standard times for each delivery,' came the frustrated reply. 'The difficulty is that with such large loads, the frequency of this type of problem is quite high and delivering an exchange machine is not credited to our allowed time for delivery. Things are getting worse. I can only cope by doing this type of visit in my spare time!'

These examples would hardly find their way on to the agenda of a board meeting. Employees raising these issues would attract a poor appraisal, a reflection that they could not do their jobs (as defined by the business) properly. The measures that probably found their way on to the agenda were more likely to have been concerned with cross-selling targets and other ways of extracting revenue from hapless customers.

In this type of sector, retail selling, the products, though important, may only be of secondary importance in relation to the whole buying experience. Few companies have managed to address this issue, but one is way out in front. Nordstrom, a retail company in the USA, has a single-sheet employee handbook which only contains one rule: 'Use your good judgement in all situations'. Some of the outcomes of this degree of staff empowerment at Nordstrom are outlined in chapter 5.

Hygiene and motivating factors

Meeting the needs of external customers, and creating a relationship that bathes them in a 'warm glow', raises the issue of hygiene and motivating factors. Many examples exist where the distinction has been unrecognised by the company, leading to unnecessary changes to process capability. Also, factors that motivate customers initially, become industry standards over time and change to being hygiene factors. Again, not recognising the transition will have companies focusing their capability in the wrong directions.

Defining the terms illustrates the difference:

Hygiene factors (aim to meet these):

- have just to be achieved;

- meeting them causes an absence of dissatisfaction;
- meeting them does not cause positive satisfaction;
- meeting them causes no contribution to differentiation;
- not meeting them causes a state of irritation, often beyond the supplier's comprehension.

For example, finding a bed in your hotel room causes little reaction, its absence causing an immediate telephone call; and choosing items on the menu that have not been available for weeks (both personal experiences).

Motivating factors (aim to exceed these):

- exceeding them leads to customer delight;
- experiencing them leads to recommendation and advocacy;
- experiencing them can outweigh a degree of poor delivery on hygiene factors;
- not experiencing them is a lost opportunity to notice differentiation.

For example, finding a bottle of champagne in your hotel room at the end of a long stay, finding a birthday card signed by all the staff presented at reception on arrival, the staff having picked out the data from the computer file (both personal experiences, which you can enjoy if you stay at a hotel called '42 The Calls' in Leeds, UK – a constant delight).

In Figure 2.5 the relationship between service level, and changes over time for each type of factor is shown. For hygiene factors, over exceeding expectations will tend to increase customer expectations in areas that are not competitive differentiators. Costs increase without any appreciable increase in customer satisfaction.

A common example is choosing 'turn-round' times as the key internal measure to achieve customer satisfaction. A building society may consistently turn round mortgage applications in three days. From the customer's perception, providing an answer within a previously agreed timescale (reliability) could be more important and relates closer to the whole house-buying process, the speed of which is not determined by the mortgage application. Getting quicker and quicker at turn-round times may eventually be counter-productive, should the customer perceive undue haste and therefore a chance that poor quality customers (potential arrears) could slip through the validation procedures and be subsidised by good customers.

Another favourite is over exceeding expectations on the delay in answering the telephone. The histogram, shown on Figure 2.6, illustrates the characteristics of dealing with callers before and after a change in policy. A characteristic of the new policy is being suddenly bounced to some other part of the organisation as another call coming through prompts the telephonists to get you off the line before a measure of their 'failure' to answer within three rings affects their appraisals. A significant number of callers experience a drop in service in terms of being put through to the right department. This impact was greater than the waiting time to get through to the company.

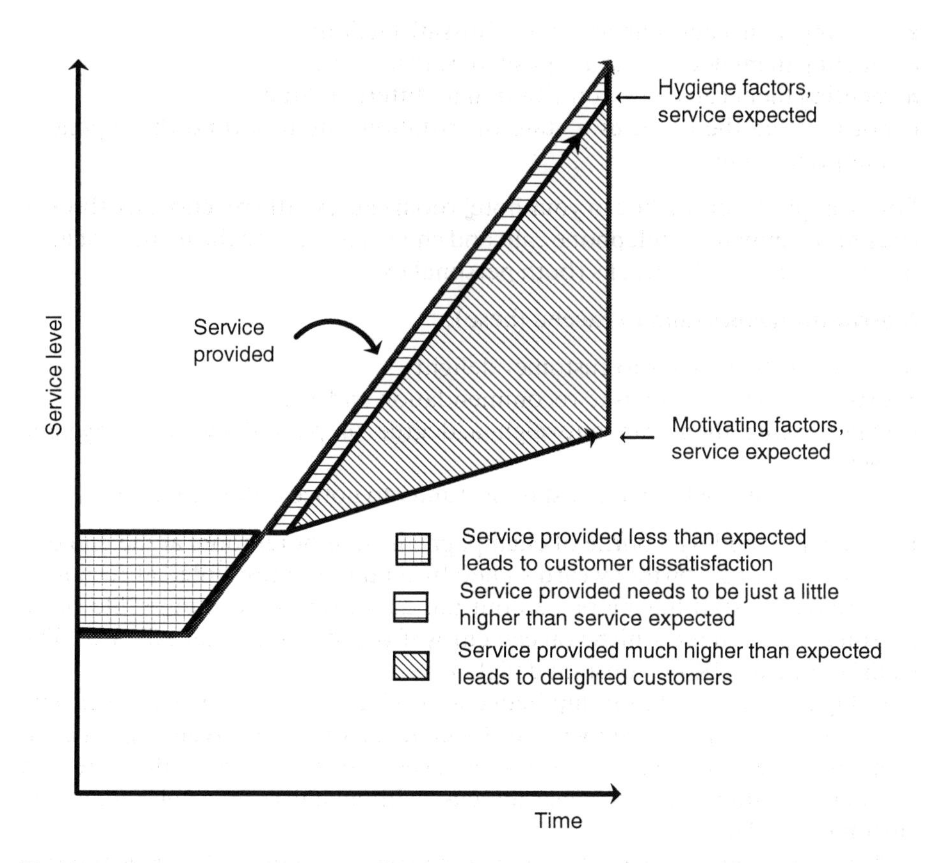

Fig. 2.5 Customer expectations

Parts of processes that deliver hygiene factors are generally easy to measure. What gets measured gets done, and becomes the key influence on behaviour. Where the process impacts on customers, then both staff and customers become victims of the company's failure to understand the influence of hygiene and motivating factors.

Motivating factors delight against the unexpected. They tend to be 'cultural' issues which arise from allowing staff the flexibility to use their judgement – the basis of trust and empowerment.

While travelling on the train, the driver announced an unscheduled stop. A passenger had got on the wrong train on his way to visit his wife in hospital and on realising his mistake, had found that the train had a sympathetic driver who 'bent the rules'. After the unscheduled stop, everyone in the carriage remarked on the unexpected event rather than complain about losing two minutes of their journey time.

On another commuter train, the driver announced at the start of the journey that he would set an alarm clock to ring over the intercom, five minutes before

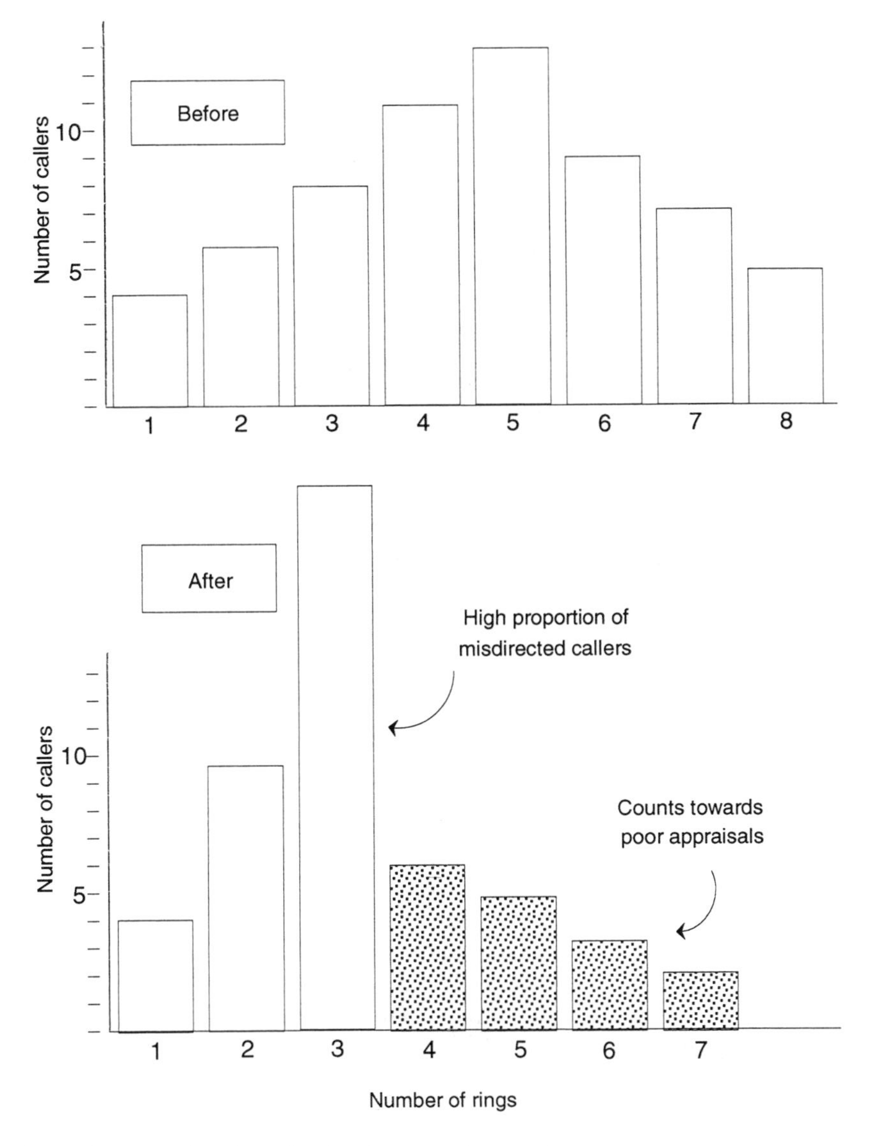

Fig. 2.6 Changing the telephonists' service level

the arrival at the first stop in an hours time. A brief demonstration followed. He also informed everyone that at the end of a long day, quite a few passengers seemed to be falling asleep and ending up at the final station, fifteen miles after the first stop and a connecting train to come back again meant a wait of forty-five minutes. No doubt there was a rule about misusing the intercom. However, right at the start of the journey, everyone remarked at the unexpected service. The atmosphere changed in the carriage from the usual silent group of dozing commuters to one where animated conversations between strangers continued for some time.

After checking out of the hotel a harassed guest informed the porter, hailing a taxi, that he believed his camera had been stolen somewhere in the hotel. He would now be unable to record his one trip to this city and that claiming the insurance money would be an ongoing nightmare. Without hesitation, the porter went and produced a sufficient sum for the guest to buy another camera and told the guest that he could pay back the sum when the insurance money came through. Two months later a cheque arrived at the hotel and a letter stating that from that moment on, no other hotel chain would get his business and, already, he had mentioned the event at an international conference.

Parts of processes that deliver motivating factors are difficult to measure and are usually avoided. No computer system can measure customer delight and advocacy, and improving customer retention takes some time before a trend will show in the statistics. Unless customers actually complain, the same difficulty arises in trying to measure customer dissatisfaction. Customers in this state are generally not so shy when it comes to telling other people stories of failures to deliver a hygiene factor and the lack of any motivating factors to compensate.

Organisation maturity

The ability of the company to change its capability will be greatly influenced by its own life-stage. Some companies will be created because someone has focused on positioning and has understood the market forces in the current environment. As a result, the new business starts with vision; a clear positioning and a clean sheet on which to build capability. Success comes from discarding the old 'rule book' and thinking through the processes from the perspective of lowest unit cost and ensuring that the product or service has an accessibility that leaves competitors behind. Such start-ups are characterised by being entrepreneurial and, to get the business going, they are often led by an entrepreneur. Such individuals and their companies demonstrate a number of characteristics:

- Management structures – tend to be radial, the focus being the visionary entrepreneur.
- Internal procedures – tend to be subjective and react to the evolving needs.
- Culture – tends to be one of great loyalty, to the company and its leader.
- People – tend to be teams and do not claim the credit as individuals.
- Finance tends to be treated as venture.
- Participation – tends to be directive, but not oppressive.
- Decisions – tend to be intuitive, creative and not hidebound by numbers.

These characteristics can be seen in brand new start-ups of new businesses, but, interestingly, the same characteristics can be seen when new divisions within established companies see a positioning opportunity that can be capitalised on only by allowing an entrepreneurial type business to start up without the entrenched and inappropriate capabilities of the parent. The growth of telephone banking is an example of this.

At a broad level, it is possible to define four organisational classifications, each of which demonstrates different potential linkages between positioning and capability, and are characterised by the main competitive stances they adopt. These are shown in Figure 2.7.

Type	**I**	**II**	**III**	**IV**
Feature	**Entrepreneurial**	**Marketing and Sales**	**Bureaucratic**	**Quality**
Structure	Radial	Divisional	Functional	Co-operative
Procedures	Subjective/ reactive	Developing	Standards	Process
Culture	Loyalty	Volume-driven	Efficiency	Customer-driven
People	Enthusiasts	Salesmen	Experts	Generalists
Equipment	Specific	Modern	Cost-justified	Flexible
Finance	Venture capital	Cash	Budgetary	Own funds
Focus	Concepts/ events	Turnover/ performance	Costs/systems	Customers/ people
Participation	Directive	Informative	Informative	Open/listening
Numeracy	Not numerate	Numerate	Numerate	Knowledge
Communications	Oral	Visual	Written	Any
Cycle	Daily	Monthly/ quarterly	Monthly	Dependent

Fig. 2.7 Organisational classifications

In a start-up company, Type I, the focus is on the entrepreneur and few rigid rules apply. Positioning is the initial driver and capability is about the flexibility to adapt quickly to any emerging need. Starting with a clean sheet, advantage can be taken of the latest emerging technologies. Here, the competitive stance will be on 'New'.

The next stage, Type II, focuses on marketing and sales in order to build rapidly or 'Steal' market share before the competition can react. If the positioning has succeeded, brand loyalty will be established and high levels of customer retention and advocacy will exist. Process capability will still be able to deliver the differentiating customer proposition that attracted customers in the first place and the processes will not be letting the business down through failures on customer-perceived hygiene factors.

As time passes, eventually, many companies slip into Type III, the bureaucratic business. In this stage they languish. Their characteristics are depressing:

- they grow ever more functionally parochial;
- they become rigorously budget-oriented;
- they focus on cost containment or reduction;
- they are rule bound, full of disempowered staff;
- they lose touch with customers and the trends in the market and competitors;
- they are neither large nor small, size is not a determining factor;
- they are not being held back by the latent abilities of their staff.

The organisation's competitive stance becomes one of 'Defend' as competitors adopt a positive differentiation. Organisations that can fall into a Type III classification are:

- those that have done well over a long time period in a generally growing market, even where waves of small fluctuations in customer demand have existed;
- those that display corporate product arrogance and convince themselves that their brand name is sufficient to overcome the increasing failures in the processes;
- those that have enjoyed a monopoly of supply;
- those that are run by subsequent generations of the family of an entrepreneur, but who lack the competences to know what to do;
- those that have achieved volume sales by targeting staff to sell in a rising market, but who cannot meet customers' true needs in a declining market;
- those that can sell because legislation states that customers must use the service;
- those where the management style is oppressive and anarchy from staff simmers daily;
- those where technological advances are perceived as passing fads;
- those where continuous learning at all levels would be seen as a sign of weakness.

The list could be longer but it is too depressing to write them down. One is reminded of too many past experiences.

Inside a Type III, a Type I or II cannot form spontaneously as it is unlikely that anyone with the right characteristics still works there. Some Type IIIs manage this only by spawning a new division under a more entrepreneurial manager with rules that differ from the rest of the business. The alert corporate board member can use this approach to achieve a cultural reverse takeover. The new division's management grow the business and absorb the parent, changing the capability and achieving corporate re-positioning. Some banks are adopting this approach.

In stark contrast, a Type IV quality company is able to respond to meet any competitive stance.

- It can bring new products and services to the market and steal market share through thoroughly understanding positioning.
- It can defend itself against competitor price discounting through constantly

being the lowest unit cost provider.
- It can nurture the existing customers through building trust and loyalty and establishing a partnership relationship with its customers.

Business Project Management is the journey to become a Type IV organisation. The current life stage of the business will determine the degree to which internal barriers exist to achieving the necessary corporate transformation. Figure 2.8 shows that the stakes are high – survival awaits the winners.

PLOTTING THE JOURNEY

BPM is the process by which companies journey along the routes of positioning and capability to deliver the vision. Depending on business circumstances, the emphasis on each axis will change. Logically, a company should start by undertaking all the activities to determine the future position, followed by changing all the processes in order to reach the chosen position.

Some companies can simply neglect to challenge and re-examine their vision of the future and their role within it. Frequently, such companies concentrate on reducing costs and increasing efficiencies, while market trends, competitor moves and the critical factors that delight customers are ignored. As a consequence, the organisation drifts and managers become increasingly frustrated at the lack of direction at the helm; decision-making abilities become stifled by the uncertainty of the goals managers are trying to achieve. The map for this journey is shown in Figure 2.9.

Each organisation needs its own map

However, it is frequently necessary to take short-term actions driven by a business imperative. In most cases, this will require action on improving the current processes, a change in capability with little change in positioning. The change in capability could be driven by a need to:

- reduce the cycle time to process customer orders;
- lower variable overhead costs;
- re-balance resources to meet current market needs;
- improve quotation times;
- reduce work-in-progress stocks;
- increase product range to meet an immediate competitor threat;
- meet changed legislation requirements;
- bring the business's capability back to the point where it is able to meet its current service promises;
- just reduce costs in order to reduce prices and retain short-term market share;
- introduce new short-term processes driven by macro-economic effects (e.g.

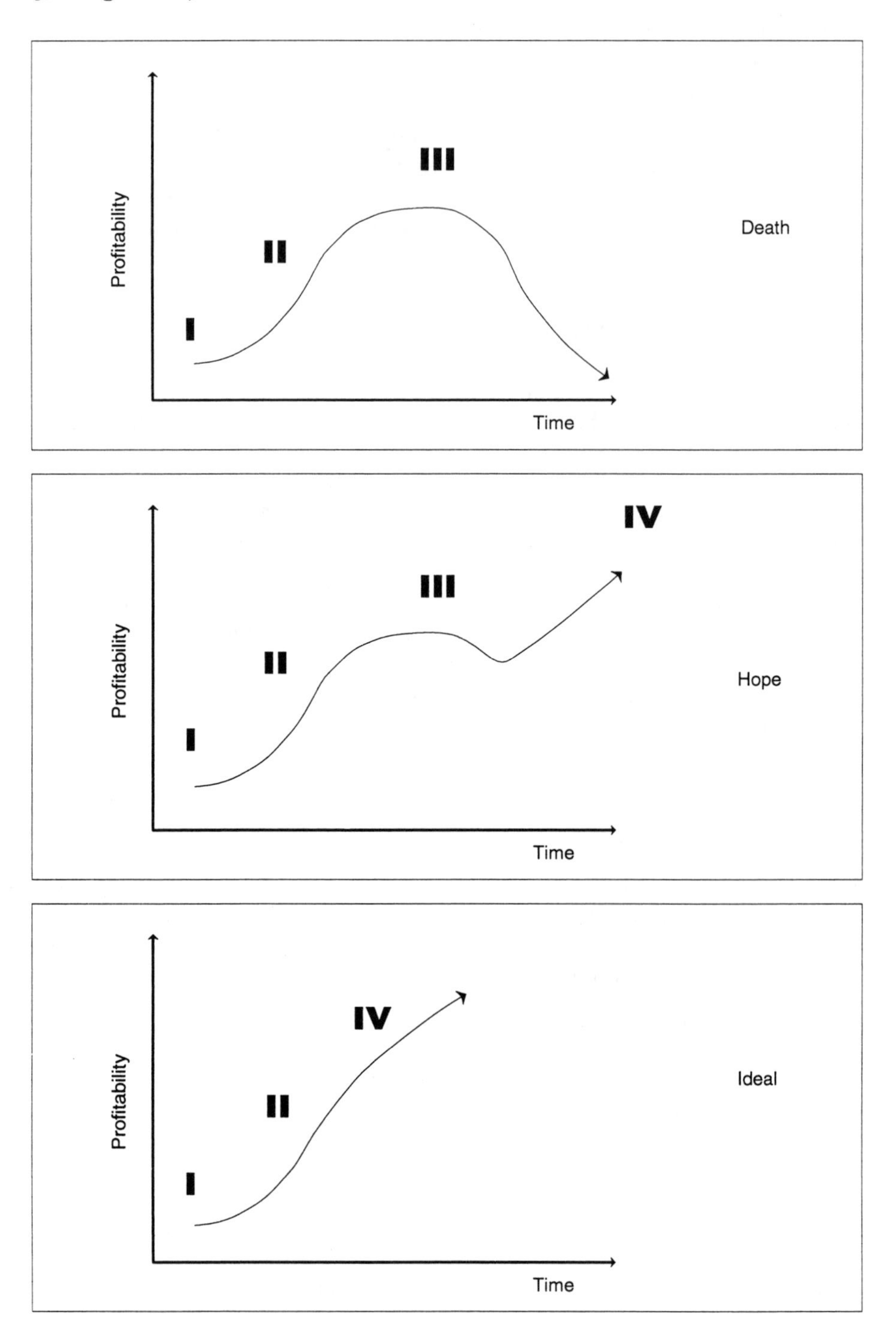

Fig. 2.8 The stakes are high

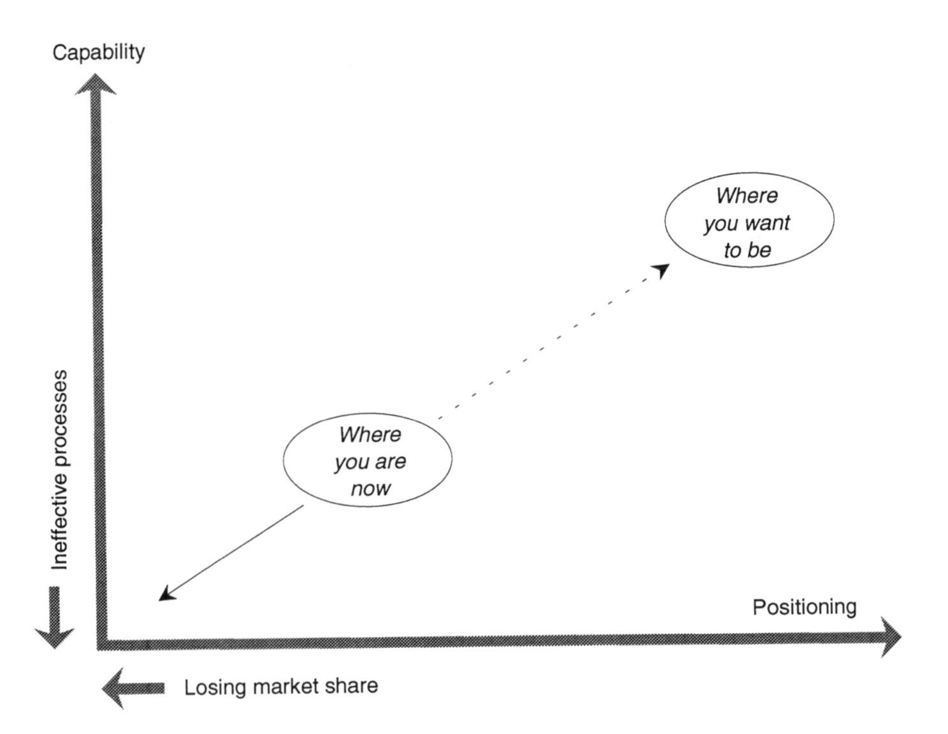

Fig. 2.9 The organisation drifts to unprofitability

increased arrears in payments from customers hit hard by a long lived recession).

Managers and staff will have a clear focus on the short-term need and an understanding of the risk of such actions on the longer-term future position. However, without the short-term action, there may be no future. The map for this journey is shown in Figure 2.10.

It is our experience that to make radical or innovatory change throughout a business from a poor base of currently failing processes is to invite failure of the attempt to achieve overall corporate transformation in one giant step. It is unwise to make a single step to the vision. Those that have attempted it have discovered that they arrive close to the vision still weighed down by the baggage of current process failure and inappropriate cultural attitudes.

The generic map, that can be amended by any business to match its unique circumstances, shows a journey that is taken in two steps:

- The first step focuses on cleaning up current processes, at which point the business will be cost effective in delivering the current customer needs.
- The second step involves radical and innovatory re-engineering to deliver clear competitive differentiation. To embark on this step requires that the organisation:

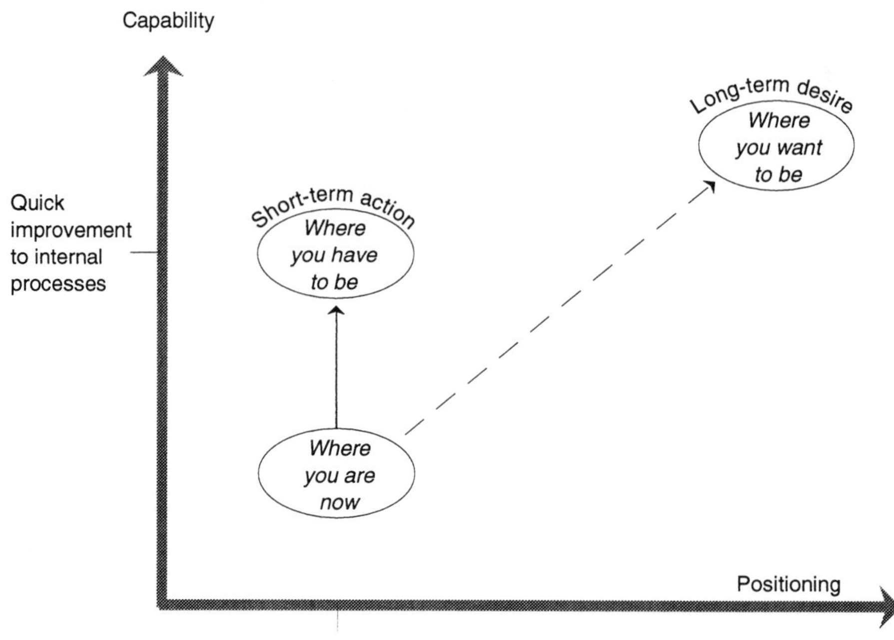

Fig. 2.10 Short-term imperative

- recognises the need to understand all aspects of positioning and has the capability to research those elements from which the future positioning is derived;
- can convert the future positioning implications into the overall corporate capability.

In Figure 2.11 the map now has specific co-ordinates and the route is plotted in two steps. The co-ordinates enable organisations to know where they are at any time, and to plot where they are going. The use of the co-ordinates is a quick way of creating a BPM nomenclature in a business and the start of a new language to reflect the beginning of the transformation.

The key coordinates on the map

The various co-ordinates on the map are defined as follows:

- **P1** (The position now)
  A description of the current business in terms of its products and markets, its type and mix of customers, its position in any competitor league tables, its current financial performance, and the current legislative and environmental constraints.

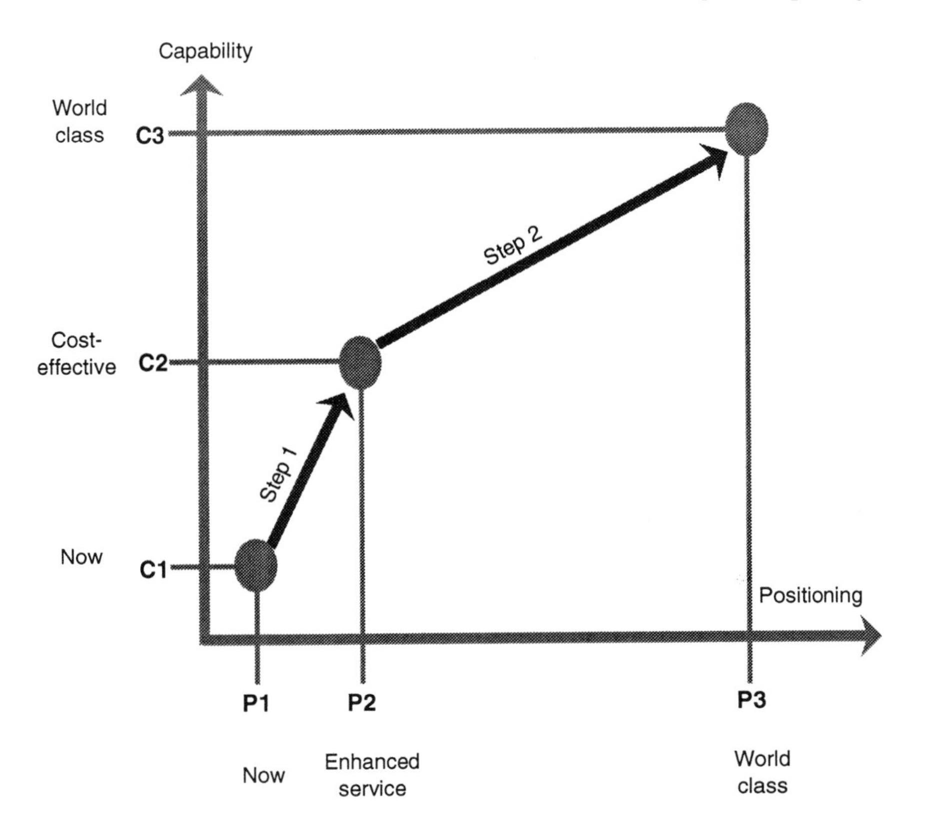

Fig. 2.11 The complete map for the BPM journey

- **C1** (The capability now)
  A description of its current way of doing business, the way it organises itself, the relationships between functions, the activities undertaken in the current processes, and the way management and staff relate to each other (its culture).

- **C2** (Becoming cost-effective)
  The changes in capability resulting from reviewing the current processes and eliminating waste. The costs in the business are reduced by improving the methods of working and by the removal of process failures in the interactions between departments, leading to continuous improvement as part of business-as-usual.

- **P2** (Achieving an enhanced service)
  The result of C2 creates a business that is more responsive to short-term changes in the market and has spare capacity to enhance the levels of service to internal and external customers.

- **P3** (World class – delighting customers, differentiating from the competition)
  A description of the business in the future in terms of its differentiating customer proposition (its DCP), the markets it has chosen to operate in, the type and mix of profitable customers it wants to retain and nurture, its position at or near the top of any competitor league tables, its strong financial performance, and the future legislative and environmental constraints it will seek to influence to its advantage.
- **C3** (World class – a lowest unit cost organisation with empowered staff and enlightened management)
  The radical and innovatory changes in capability to deliver the future position (P3), resulting from creating a series of key business processes aligned to obtain customer retention and advocacy. The processes are in an internal framework that continually delivers at lowest unit cost, with management and staff relating to each other in a cultural framework of mutual help and team-work leading to continuous innovation.

Taking the first steps on the journey

As has been outlined above, companies can set off on the journey taking different routes. If the company is starting from a point that has the business poorly positioned to meet the challenges of the future (P3), and is letting current customer service (P1) deteriorate through perpetuating ineffective current processes (C1), then it will need to make the whole journey.

Some actions can start in parallel, others will be consecutive. The sequence would then be:

1. Define the point, P1/C1.
2. Through customer needs surveys or similar methods, establish the shortfall in current external service to close the gap to meet current customer needs, P2.
3. Start the programme that will improve the current business processes to eliminate waste and start to align the business to processes.
4. Start the programme of research and understanding that will provide the knowledge that is P3.
5. Implement the changes that emerge to achieve a new capability, C2.
6. Introduce additional tools and techniques of continuous improvement to achieve C2+, C2++, and provide further help and guidance to align management behaviour to process thinking, team-working and helping staff.
7. Close the gaps in everyone's range of competences to enable everyone to do the jobs that are required of them.
8. Summarise the conclusions from the P3 work and outline the implications on capability, C3. This stage points to the fundamental need for corporate transformation.

9. Deploy the need to change to P3/C3 to everyone in the organisation.
10. Using a clean-sheet approach, work through the detail of C3, using an iterative process that involves managers and staff in the radical and innovatory changes that will be emerging.
11. Implement the changes to achieve C3.
12. Provide further help and guidance to align management and staff behaviour to continually delivering the differentiating customer proposition.
13. Close the gaps in everyone's range of competences to enable everyone to do the jobs that are expected of them to deliver the differentiating customer proposition.
14. At any time, define the new point P1/C1 and repeat the cycle from (1) to (14).

The rest of the book will follow the sequence listed above, as it represents the complete journey. Each organisation will need to decide for itself where it is on the map and start the journey from that point, taking the route that best suits its own current circumstances. It is worth repeating the old proverbs that opened this chapter: *If you don't know where you are going, any road will take you there. If you don't know where you are, a map will not help you.*

3

THE BPM JOURNEY: Step 1 – Improving current processes

We sometimes forget the meaning of the word 'company', people who co-operate with each other to deliver products and services to others.

Nick Develin

INVOLVING EVERYONE IN IMPROVING CURRENT PROCESSES

A company's major asset is its people. The people manage and work in the processes and know when they are going right and when they are going wrong. The people may not know all the causes of their problems but they will be spending time on dealing with the symptoms. Their time is valuable, they want to do a good job but often the process defeats them. Through their involvement in improving the current processes, the organisation gains spare capacity to re-allocate resources to improve the relationship with customers. Through involvement, pride in workmanship, at all levels, is restored. Through involvement, commitment is gained and implementation of change accepted.

The starting point – P1/C1

For your own historical archives, we would suggest you describe this point, so in future years everyone can look back with nostalgia at the bad old days. At points along the journey the incremental change may be frustratingly small from day to day. Looking back to the point C1/P1 will show you just how far you have gone. By pulling together data from around the business a description of the current business can be put together. This is best undertaken at a board members' workshop in order to start with top-level commitment to the BPM process. The behaviour of board members will also begin to highlight where total or just begrudging commitment lies and the degree to which acceptance of the current situation is seen as an acceptable strategy for the future.

In overall terms, the current business may have the following characteristics:

- Financials – (e.g. low performance).
- Competitive position – (e.g. near bottom of table).

- Products and services – (e.g. product/volume-driven).
- Structure – (e.g. functional and parochial).
- Management style – (e.g. control/lack of trust/striving/win-lose, etc).

The following excerpt from a finance company's C1/P1 description indicates some of the stark realities that companies may be facing.

> In the past, our own and our competitor strategies were all similar and relied on treating the customer as an asset – an asset to be milked. Net income margins, cost–income ratios and asset growth were calculated in advance, from which all other targets were calculated. This essentially works in a rising market. The equations have broken down in a constantly falling market as revenues decline and this has been exacerbated by our failure to keep up with competitor enhancements in customer service. The focus has now switched to a cost–income ratio. Our competitors enjoy better ratios than ourselves and are getting better. Hence, competitors will earn greater profits per unit of income or, more importantly, can afford greater inducements to lure customers toward them. The equivalent of a price war may be good from a customer's perspective, but if traditional methods to grow market share by buying it are used, then the knock-on effect to ourselves and others leads to a shake-out in the market.
>
> At the moment, severe cost control is the only way to achieve the necessary cost–income ratio with absolute management expenses remaining static. We know from benchmarking our competitors that they have more efficient processes than ourselves, and that they operate in a less functionally parochial structure, with fewer layers of management. We have no choice in the short term but to make cost reductions, even though we know that such action could put at risk our ability to secure longer-term income. . . .

In this case it was necessary to make the first step in the BPM journey to the point C2/P2 on the map.

Primary objective of Step 1

The primary objective of Step 1 is to examine all or part of a company's processes and to develop implementation plans to improve efficiency and effectiveness. In overall terms, the key drivers for change could be that costs must reduce while simultaneously improving quality and current customer service. Step 1 provides a burst of activity that brings out the major proposals for process improvements and uncovers many minor irritations due to local process failures. Implementation of the proposals from Step 1 needs to be absorbed into the transition to an ongoing state of continuous improvement.

The key deliverables of Step 1 are focused on the identification of an interim workable set of proposals for change, given the existing environment and level of change that can realistically be achieved in the short term. This delivers the changed capability, C2.

Cost reduction

A key deliverable of Step 1 may need to be a materially reduced operating cost. At the moment, the current service from any department or section could be excellent, good, bad or indifferent. While maintaining this current level of service, the level of activity could be reduced through improvements in the methods and processes in a department, and through the elimination of another department's errors and failures that passed problems through the processes. However, because some current levels of service may be too high or inappropriate for the current business needs, a reduced level of activity needs to be explored by proposals for selective reduction below the current level of service. Such proposals force the debate on whether this creates real risks to the business or not.

Service enhancement

Another key deliverable may need to be an improved quality of products and customer service through additional activity to enhance the service from certain departments above the current level, or, through new tasks, to provide outputs which the business does not benefit from currently. In some departments, an increased level of activity or service may be necessary in one department in order to achieve outputs which are right-first-time and thus save time in other departments who are downstream in the process.

Re-balanced resources

In general, for the scope, the business could be seeking to re-balance the resources with an overall cost reduction and selective increases in service to provide additional benefits to the business greater than the cost of providing them. Figure 3.1 shows the actual outcome from a number of companies implementing Step 1. The vertical bars are the number of departments or sections within the scope that were required to implement a particular percentage change in resource level. Sections reducing by more than twenty per cent were generally those where the level of service was to be reduced. Those between ten per cent and twenty per cent reduction were usually those remaining at the current service. The shape of such a graph is unique to each company and represents the appropriate overall level of service determined by the company at the level of costs it can afford at the moment, in order to maximise its business in the current competitive and economic environment.

Choosing the scope

A range of possible scopes is shown in Figure 3.2. In embarking on Step 1, the best results are obtained if the company does not place any preconceived conditions or constraints on the possible outcomes. However, where other

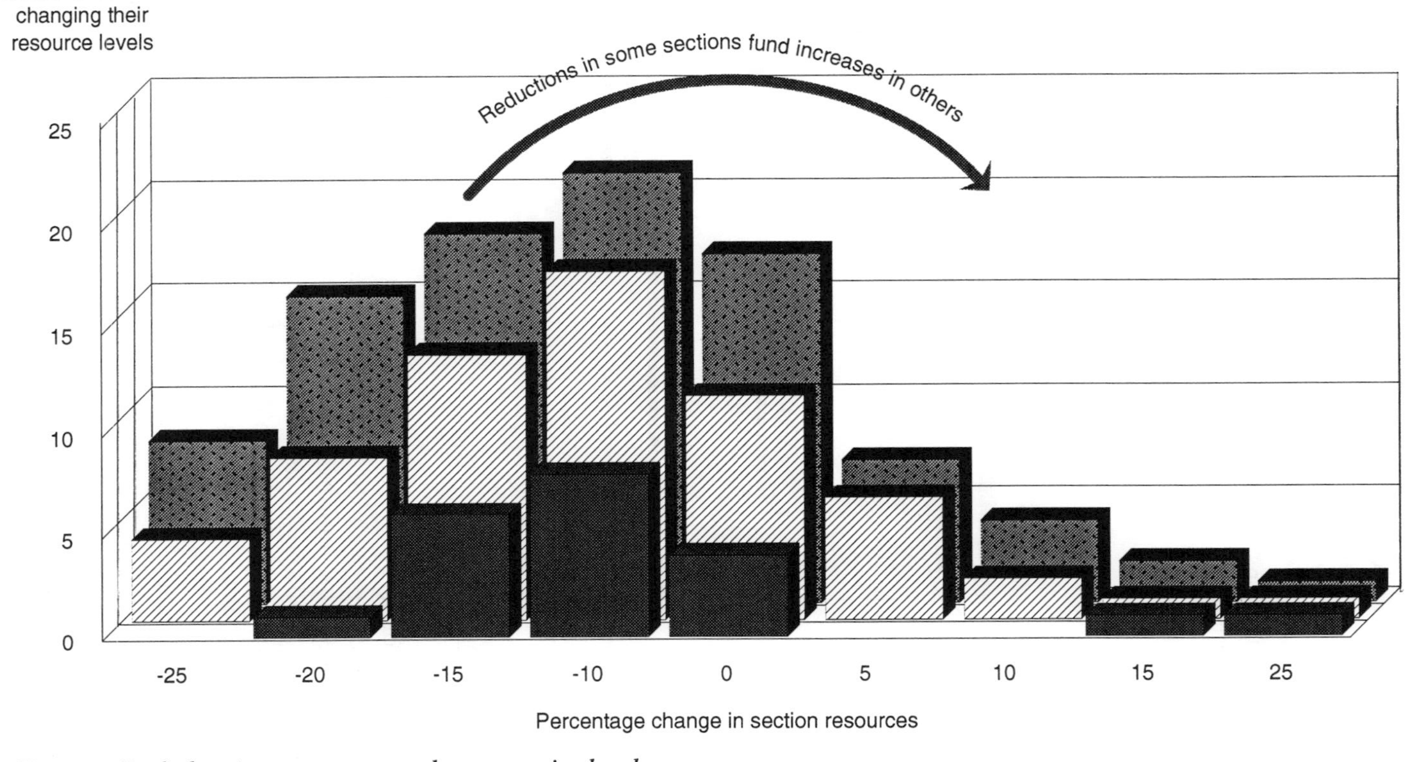

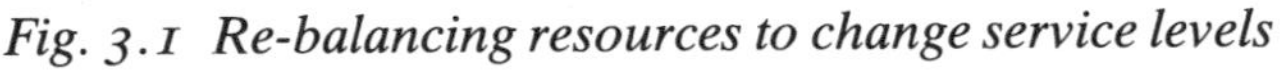

Fig. 3.1 Re-balancing resources to change service levels

major initiatives have already started within the company, then these are normally considered within the overall BPM umbrella. In nearly all cases, companies will have undergone change in the recent past prior to starting on BPM. There are no pre-conditions to starting Step 1 although it will be necessary to be aware of any other initiatives that have already been started within or across functions. It will also be necessary to position the need to start on Step 1 in the context of the company's overall plans for the future, particularly when communicating the start of Step 1 to all of the employees.

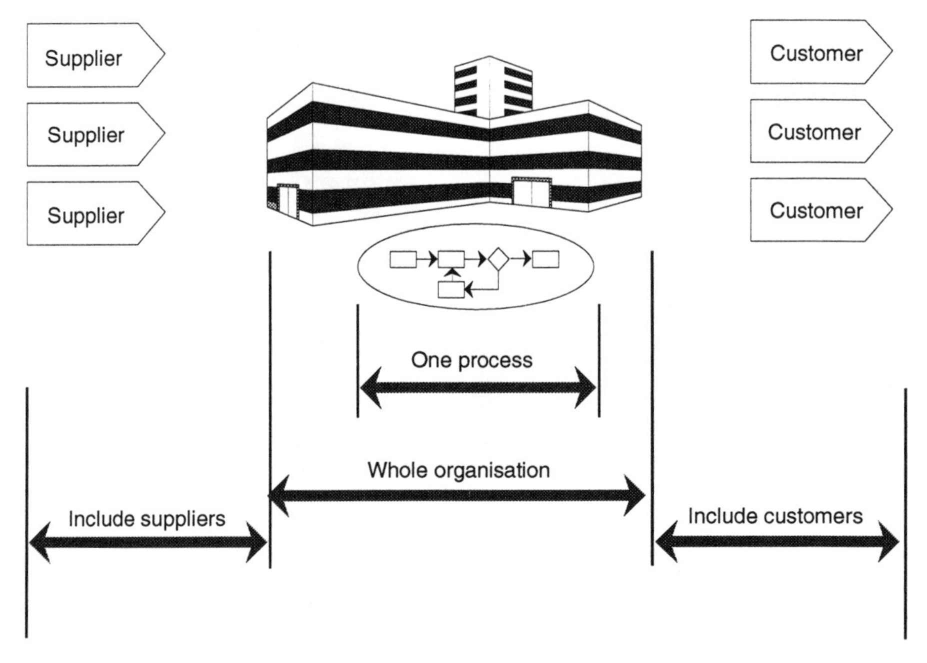

Fig. 3.2 Choosing the scope

As the purpose of the programme is to examine how resources are used in the current business processes and to allow everyone to develop ideas for changing the way things are done, improving all processes and enhancing quality, then clearly the success of the programme relies totally on employee input. The role of any facilitator or support team should not be to develop ideas on the staff's behalf, but to stimulate and encourage an atmosphere within which constructive suggestions can be made. It is important that staff own what is produced at every stage of Step 1 in order to achieve commitment to the emerging proposals for change and to move towards continuous improvement.

A fundamental requirement for success is the notion of internal customers and suppliers working together in cross-functional processes. Although best results are obtained when Step 1 is company-wide, in some cases, companies begin with a more limited scope in the form of a pilot study. A key feature of

BPM is the emphasis on exposing the nature of internal customer and internal supplier links within the business. Where the scope is less than the whole company, Step I will still need to involve staff outside the scope in order to review any proposals that affect them from staff within the scope.

A number of emerging issues will impact external customers and external suppliers. Where Step I identifies that an external view should be sought, then this will need to be considered within the company in terms of the impact this could create with external customers and suppliers. Where the views and needs of external customers are not known, then Step I will need to include some type of customer needs survey. This is discussed later. Where external suppliers' processes are found to be contaminating one's own business, then part of the implementation of change will require a greater involvement by external suppliers in the change process. Where an external supplier's processes are heavily integrated with your own, then the scope of Step I can involve the supplier to the benefit of both businesses.

Facilitation support during Step I

Embarking on a programme of improvement that involves all employees needs to:

- find time inside the already busy days of everyone;
- move forward in a consistent and disciplined manner.

A support team, skilled in the techniques of facilitation, overcomes these difficulties and provides a central point where data and insights can be cross-fertilised and spread wider across the whole organisation. Finding people from within the organisation to provide this team support will itself create some additional workload in the areas from which they are drawn. However, the use of the support team saves much time later throughout the organisation and reduces the overall lead time to achieve improvements.

Members of the support team also find the role very rewarding, in career terms, as no line position affords the opportunity to see so many cross-functional issues in a short period of time. The organisation should therefore see the use of team members as a key investment in management development for the people that are chosen for this role.

The framework of support

Ideally, the stages within Step I should be structured in order to avoid a random series of initiatives being generated, ranging from sceptical non-involvement through to the over-enthusiastic generation of proposals to change everything. A structured approach requires that people within the organisation perform a number of key roles during Step I. These roles are shown in Figure 3.3. To co-ordinate all the stages in Step I and to provide facilitation of the change process, a small team of facilitators, drawn from all levels within the business, will need to be seconded full-time on to the programme. The facilitators each look after a number of groups of employees. Their role is to assist the staff in the scope, but they should not be preparing any documentation or proposals for change on behalf of the staff.

The team of facilitators will need to develop the appropriate methodology to suit the organisation. The methodology outlined in this book is generic in that the key elements will be appropriate for most organisations.

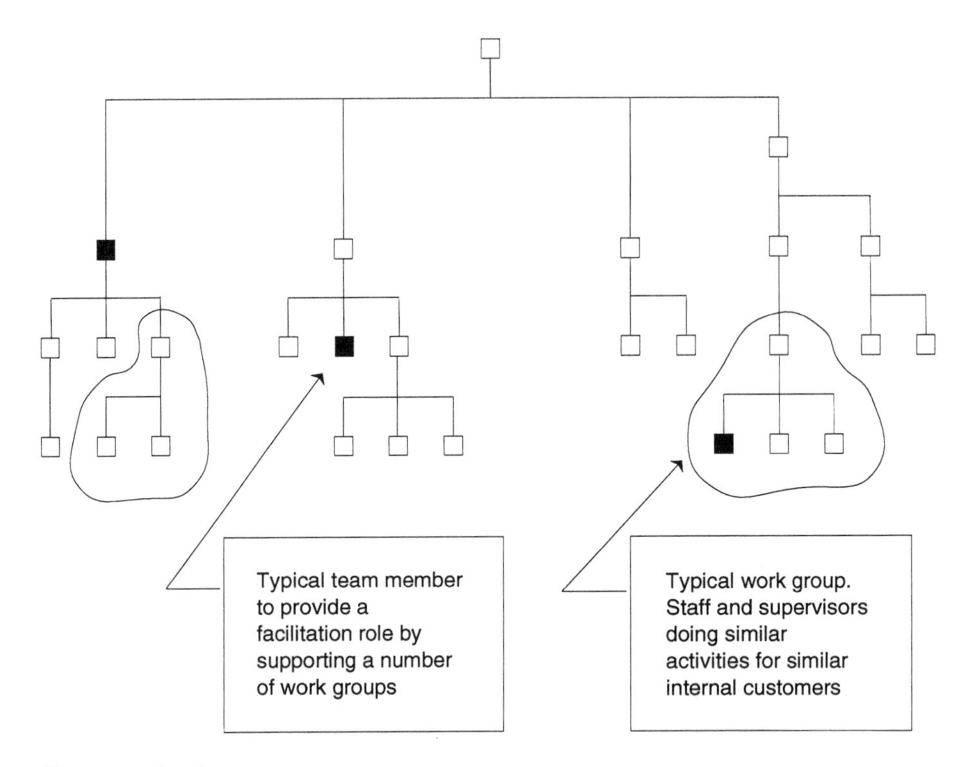

Fig. 3.3 Facilitation and support in Step 1

For the purposes of Step 1, the organisation should be split up into small groups, where the members of each group will have broadly similar objectives and activities. These groups will normally follow the current structure with each group generally being the equivalent of a section within a department. For most companies, an average of ten staff will be in each group, although some groups can be as small as one person where their activities are of a specialist nature. Each group should nominate a leader for the duration of Step 1. This person need not be the current organisational leader, but the nominee needs to have a reasonable level of experience of the group's working practices and the nature of the process links to other groups.

The work group leader's key role will be the individual through whom any support team members can make contact with the group. The leader will also have the task of being responsible for the creation of a set of data, on behalf of the people in the group, by co-ordinating the collection of the data from the people within the group. The leader will also ensure that any key deadline dates are met throughout Step 1, and co-ordinate these with the support facilitator. Data from all groups, analysed centrally by the support team, can be communicated from the team through the leader to the people in the group.

A timetable will need to be applied to ensure that the whole programme moves forward at the same pace throughout the areas in the scope, to ensure that proposals for changing a multi-functional process are being considered at the same time by those groups working in the process. Each facilitator should be able to support around ten groups and best results are obtained where the facilitator from one function supports groups from other functions. This avoids the problem of facilitators representing a function and bringing entrenched parochial influences to bear on the proposals for change.

Communications

Starting on Step 1 without fully briefing all the employees will doom the initiative to failure. Where the organisation is carrying some cultural baggage from the past, then the normal communications methods may well be inappropriate, both in terms of mechanism and style. However, the initial communication will signal the cultural and business framework for the programme and can therefore be used to demonstrate the first changes in the senior executive attitudes, beliefs and behaviours. Above all else, the essential requirement is for honesty. If Step 1 is to address a business imperative, such as rapid downsizing involving a reduction in the number of employees, then this issue should not be hidden. Employees can take bad news as long as they perceive that management are in control of the situation, and that the future will have some positive certainty after Step 1.

Statements from the executive group that imply that the current poor business performance is entirely down to the poor efforts of the employees only affirm that a blame culture exists and any ideas for improvement will be seen as 'exposure' of the poor employees and managers. Such a stance will discourage anyone from contributing positively.

Similarly, exhortations to do better and work 'right-first-time' also signal where the cause of problems is seen to lie. If employees have been crying out for help from management for some time, and their cries for help have been consistently ignored, then they will not see Step 1 as being any different from business-as-usual. The dilemma for some organisations is one of being honest and credible when no previous track record for this style exists. Even an honest communication will be viewed with suspicion. There will be no other option than to start Step 1 and demonstrate throughout the programme that the executives have been and still are honest, and thus build credibility as time passes.

ESTABLISHING THE CURRENT CAPABILITY

Data collection and analysis is a critical first stage in Step 1 as it provides an objective basis for:

- understanding the current failure to meet current external customers' needs;
- challenging the existing output or levels of service, both externally and internally;
- evaluating the benefits of improvements of method;
- understanding cross-functional organisational relationships;
- understanding the interactions within multi-functional processes;
- identifying and evaluating systems opportunities.

The data need to provide the basis for rational, fair, objective and open discussion and decision-making about improving the efficiency and effectiveness of the business in the short term. The facilitators work closely with each group to create a set of data. These data should include organisation structure, organisation and process affinities, problems with internal services received or provided, activities carried out within each group, the elapsed time of activities, and their classification to reflect their level of added value or otherwise. Process mapping is also key to tracking the cross-functional flow inside a business process. Such maps quickly highlight failure feedback loops and potential over-complication within a process.

Insights from data analysis

In Step 1, data collection is the 'Voice of the Process'. The range of data is unlikely to be already collected routinely within the organisation. The data should include an understanding of current customers' needs, the nature of the flow of the processes within and between functions, the perceived level of service given and received, and the type of activities undertaken within each process. Simple analyses allow powerful insights into the business to be obtained quickly and allow the processes to speak intelligently.

Customer needs surveys (external)

Without asking customers, companies should never presume they are meeting the current customers' current needs. If the evidence is missing, then Step 1 will need to begin with some involvement of external customers. The exact nature of their involvement will depend on the nature of the business. For a High Street branch outlet business the survey could involve questioning customers as they leave the shop. For a finance company the survey could be a postal questionnaire matched to internal data on segmentation. For an airline it could be a comparative analysis with other airline companies on the same route, questioning regular flyers on a range of airlines. For a design company it could be a discussion panel of customers across a range of household types.

Wherever possible, obtaining information on a competitor's performance relative to one's own provides an insight into the degree of differentiation you or your competitors may be achieving. Figure 3.4 shows the results of a survey undertaken by an office equipment and stationery supplier. The list of variables were created by the customers without prompting. The total list of different factors mentioned came to forty, although only the top twelve most important are shown in the Figure.

The diagram lists the factors in descending order of importance. The customers were also asked to indicate whether the particular supplier was perceived as being worse, the same, or better than the competition. A classic 'X' diagram appeared, the most important factors to the customers attracting a score on performance of achieving the worst level of service relative to other suppliers. The situation was reversed at the bottom of the diagram. It was interesting to note that 'Service staff' fared rather better than the trend. In reality, the service staff were overcoming the difficulties posed to customers by slow delivery and poor availability. Luckily, the service staff were providing the company with a reasonable reputation which stemmed from their dedication to solving the problems caused by the failures in their own company and the service staff's high levels of sympathy with the customers' predicament.

By scrambling the list of factors and asking the supplier's management to rank the factors in what they perceived to be the relative importance to customers, then the root cause of most of the problems became apparent. The supplier's priority was considerably out of line with the customers' needs, but the processes and staff in the company were being focused and managed on the basis of their own management's priority. Good performance, brand loyalty, customer service staff and buyer inertia accounted for the reasonable level of business performance. However, the business was significantly at risk from the changing perceptions of levels of service as the primary evaluator, particularly as customers began to operate on low levels of their own stock.

Meeting external customer needs – a word of caution

- A firm sold its products through a chain of third-party distributors. The survey indicated that its immediate customers, the distributors, wanted high levels of service on deliveries – next day delivery rather than a weekly top-up. When this had been achieved, the supplier found that it became inundated with many small orders on a daily basis. Further investigation showed that the distributors simply de-stocked and used the additional space to carry a wider range of competitor products.

  In an attempt to overcome this problem, the supplier offered price discounts for larger quantities and also installed computer terminals at the distributors linked to the supplier's current stock file, the idea being to allow the distributors to know whether to offer alternatives in the range to

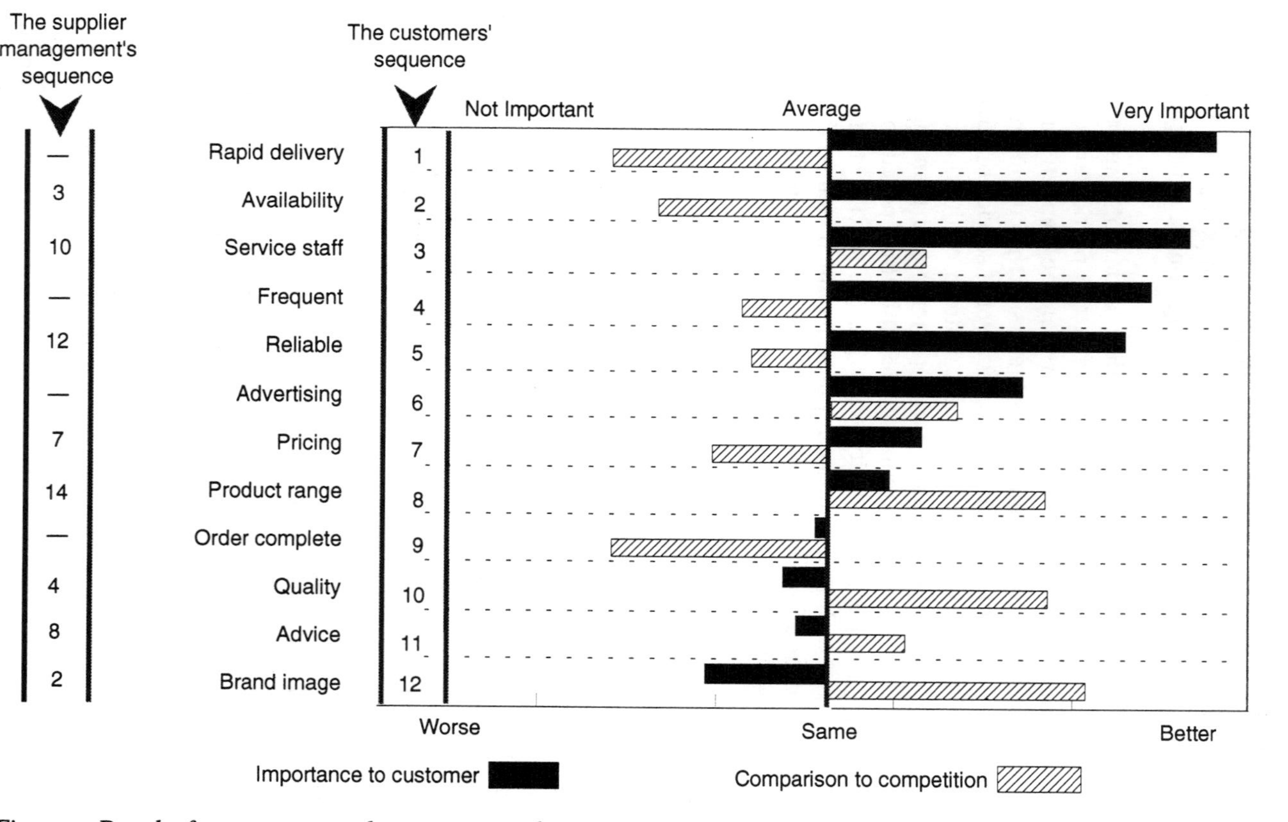

Fig. 3.4 Results from an external customer needs survey

customers or order from the supplier's stock. Curious at the frequency that a large range of part numbers were being interrogated on the database, the supplier found out that their competitors' sales representatives were using the terminal on the distributors' counters to interrogate the supplier's stock file and then offer their own products where stocks were shown to be low or at zero.

- A bus manufacturing company had traditionally met its customers' needs by supplying vehicles that were robust, fuel efficient, easy to maintain and internally hard-wearing. With a change in legislation (bus deregulation), a number of bus operators could now use the same routes competing for the same passengers. The bus operator's customers were now making choices based on a number of new factors such as comfort, warmth, safety, and so on. The bus manufacturing company initially failed to appreciate that the travelling passenger was now the key customer to influence on the basis that the operators were now influenced by passenger choice.

  By building prototypes of a superior design, based on surveys of the travelling public, and allowing operators to trial the vehicles on routes where they faced stiff competition, the manufacturer was able to prove that not only did the vehicles increase market share, but also they attracted more car drivers to become bus travellers – an investment that paid off, once the right customers had been identified.

- Questionnaire design is important. The supplier can fall victim to the cultural problems it will need to overcome later. Leading questions based on the supplier's own prejudices creates reinforcement of the current problems. To the question 'would you like our switchboard to answer in one, three, or any number of rings?', is likely to prompt an answer of, say, three rings, but the questionnaire would not have prompted the real drivers of customer choice. It is also unlikely that a hotel will directly ask a question of the type, 'if our staff were empowered to spend any amount of money on customers, would you find this desirable?', as management would probably not want to see a 'yes' as the answer.

In general, questionnaires deal with hygiene factors. Knowledge of motivating factors comes from a broader knowledge of many sectors and industries and an understanding of the overall trends and influences on business customers and the general public. Where the expectation of customers is low for a whole industry, then the customers are unlikely even to think of any motivating factors. However, it is just this which provides the opportunity to delight and differentiate.

Customer complaints

Customer complaints are a form of unsolicited customer needs survey. While wishing to have a business that does not generate complaints, in the short term, complaints analysis is a useful mechanism to determine where some priority should be given to improve processes.

From research undertaken by TARP, a company specialising in undertaking complaints research, on average a dissatisfied customer will tell twelve others. The problem for the service sector, in particular, is that for many companies, they are in an undifferentiated and competitive market, but repeat business is vital to sustain profitability. However, it is in this sector that people tend not to complain. Typically, they do not think it will be worth the time and trouble, they do not know how to go about complaining, and they have a fear of retribution should they need the service again. One should always assume that the level of complaints actually received considerably understates the overall perception customers have.

The data from research undertaken by TARP suggests that the process of handling complaints, while they exist, is as important as getting the core processes performed right first time. Against expectation, completely satisfying a complaining customer is itself a motivating factor.

In an example from an insurance company, the complaints department were largely ignored and the process of handling complaints became simply a routine procedure. Letters of complaint arrived and were filed for action in date sequence. Progress on letters was slow as telephone complaints physically intervened and took priority. Frustrated customers who wrote again to chase their problem had their file removed, their next letter added to the file, and the file returned to the back of the queue. At a critical rate of writing, a complaint would never be answered. The number of complaints had doubled from 4,000 to 8,000 per year over the previous three-year period. Clearly, a crisis was just around the corner.

Benchmarking

Benchmarking is a technique to compare a variable, output or process in one's own company to that achieved by others. It can be used to focus on those competitors which, by accepted sector standards, are more successful than one's own business, and then to focus on those aspects within the competitors' businesses that are providing their success. For example, in a comparison of the hours required to build motor vehicles, Japanese manufacturers consistently had productivity levels that outstripped their USA and European counterparts. As this reflects on product prices then this knowledge would focus competitors on discovering how to achieve and outstrip Japanese levels of productivity.

In the finance sector, cost–income ratio could be a prime indicator of success and one's relative ranking in a league table of these ratios would provide a spur to attempts to understand the differences in each company's operating methods. At one level, cost–income ratio may be a prime measure used by the city to establish a credit rating for wholesale funding. However, at another level, cost–income ratios may be distorted in any one year as costs rise to put in place actions to increase significantly revenues in another year. High-level comparisons may not always tell the complete story.

In a non-competitive situation, benchmarking provides a comparison between processes. Level of inventory may be an indicator of better stock management processes, number of debtor-days may be an indicator of a better process for payment collections or it may be an indication of a better type of customer relationship or a different type of customer segment. For any benchmarking exercise to be meaningful requires more knowledge of one's own and the benchmarked companies than just the variable being measured. A range of approaches to benchmarking is shown in Figure 3.5.

Method	*Application*	*Pros/cons*
Internal comparison	Multi-site/branch	• Low risk/low reward • Can excellence be achieved?
Co-operative ventures	Where comparable sources available	• Difficult to gain partners • Differing goals • Need to share data
Third-party surveys	To establish current positioning	• Lack of detail • Truthful?
Customer surveys	All, various methods	• Can you deliver?
External standards	Generally TQM oriented (European Quality Award, ISO 9000)	• Are they the best? • Focus on your needs?
Cross-industry	High knowledge and focus on key processes required	• Low threat • Excellence in specific areas • Mutual gain?

Fig. 3.5 Approaches to benchmarking

The two key difficulties are finding co-operative companies with whom to benchmark and then ensuring that you are comparing like with like. It is also important to measure variables that are important to your business in relation to improving aspects of customer service which are relevant to your customers and those processes that have a high potential for cost reduction. The risk is that whole areas of benchmarking expertise grow within the company in every function and becomes an industry within itself. Anything that can be measured is benchmarked until the company disappears under a mountain of irrelevant comparisons.

Benchmarking is not an end in itself. Its use should concentrate on:

- helping a company focus on what is important in its competitive environment;
- providing a framework for systematic analysis and learning, and a reduction in myopia;
- identifying what is possible and achievable;
- acting as a spur for change; and
- encouraging team-work during the benchmarking exercise.

Customer needs surveys (internal)

Internal customer needs can be surveyed in order to discover how the services supplied match the needs of the internal customer in relation to the perceived importance and performance. In an example where a Publication department served a Marketing department, the results of such a survey highlighted a mismatch between the service provided and that required. Only the supply of the annual report, a company need, was marked as a service higher than required. All other important needs, such as liaison with sponsors, video creation and printing, tended to be wholly inadequate. Later analysis of the Publication department's time showed how little was devoted to meeting the needs of the internal customer. Most of the time went on 'other' activities, those activities that the Publications department had decided for itself were important without reference to any internal customer.

Internal surveys at the start of Step 1 are enlightening for those involved, and can create a fair degree of emotion if they raise old interdepartmental antagonisms. However, the survey is an early thermometer for gauging the heat that could be generated when internal customers and suppliers get together later in Step 1 to discuss process improvements. Some heat is healthy at this stage as it focuses on the inevitable clash between meeting functional rather than process objectives, and will demonstrate how naturally people blame others rather than look first at the processes in which both work.

Meeting internal customer needs – a word of caution

At the start of Step 1, any functional parochialism that exists will be reflected in an internal customer needs survey. Any department that is busy meeting its own objectives out of line with those of the business will be demanding services from others to perpetuate its own objectives. Where this situation is known by the supplying departments then there will be little motivation to adjust or improve their service to meet unnecessary internal customer needs. This type of issue is dealt with in later stages of Step 1, when the overall processes will need to align with the business objectives. Again, some heat raised by this type of issue provides early signs of the behaviour of some departments and their managers. At this stage, give everyone the benefit of the doubt and assume that such behaviour is generated by ignorance of the bigger picture rather than being motivated purely by self-interest. Everyone needs the chance to learn that a change in behaviours as well as processes is a potential outcome of Step 1. The support teams often suffer a certain level of frustration when they are confronted by such poor behaviours.

Process flow charts

In conventional linear process charts, activities, decision points and so on, are drawn one after the other in a vertical column. Such charts are often used to analyse systems with a view to computerisation. In BPM the emphasis is to understand the interaction between the people and the processes in which they work, particularly the points where the process crosses functional boundaries. Such flow charts put the sections, departments or functions across the horizontal axis and the process is charted as it moves from one department to another. We use the term 'the cast of characters' to represent the people in the process.

For most processes, using just rectangles to indicate activities and diamonds to indicate decisions or check points is sufficient to generate an understanding of the process. In Figure 3.6 we see a simple purchasing process from a government research establishment. Even at this level of detail the staff involved became quite animated concerning the issues it highlighted. In this example, both 'Technical vetting' and 'Purchasing' had assumed the right to challenge the requirements of the requisitioner, the scientist who held the budget to undertake the research. 'Technical safety' also insisted that every electrical item delivered be taken apart and re-built to check for electrical safety, a procedure that itself created risks to the end-user of the equipment.

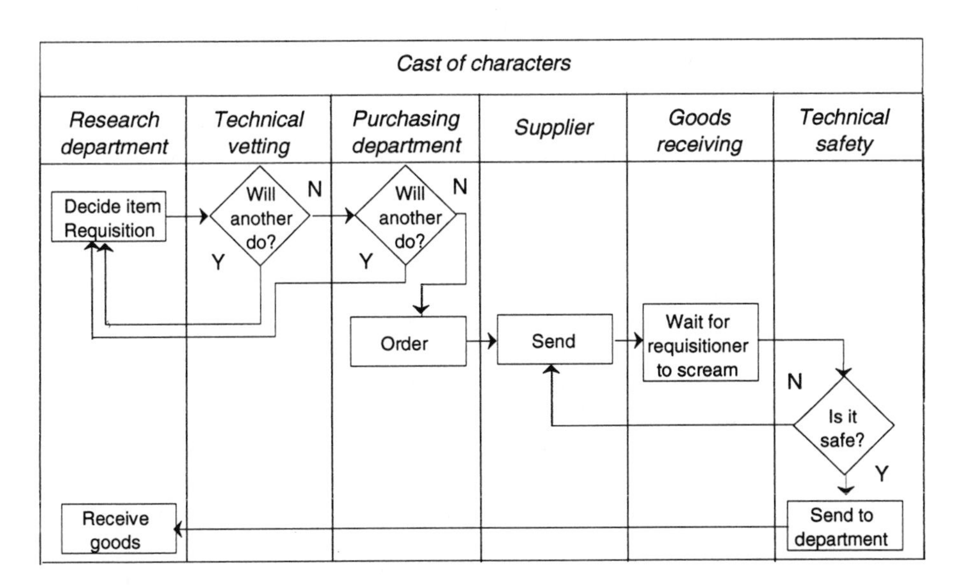

Fig. 3.6 Frustrations in a purchasing process

A process chart will start to beg further questions. Each horizontal line is an internal customer/supplier relationship. At this point, the process can be measured to understand the nature of the service given and received. It is interesting to note that when an external supplier receives an order, at that point the external supplier is the customer. If the supplier cannot understand the exact requirements on the order documentation then there is a high chance the requirements will not be met. Anyone who has been on the receiving end of a fifty-page 'request-to-tender' will have sympathy with this problem, although pointing out the real nature of the relationship at that point with the buying department is often to invite a contrary view.

Each decision or check-point tends to have a feedback loop. Such a loop indicates a failure to provide a correct output from the upstream activity, resulting in checking and re-work. By measuring the frequency by type of error, the priority to search for root causes can be established.

After a thorough review of the process the roles of each department changed. 'Purchasing' changed to providing a quality-assured supply role and setting up global contractual relationships with key suppliers from whom the researchers ordered direct to their specification. The 'Technical vetting' and 'Technical safety' roles disappeared, although the expertise was re-used in support of the researchers as part of the supplier validation process. The support was on request rather than provided irrespective of need.

Another insight that process charting provides is the complexity of reality. When drawing a process chart, it is quite common just to draw how the process should work rather than how it actually works. Figure 3.7 could be a typical supermarket. When all is going well, the process is quite simple and this could well reflect how the shop manager believes things are happening all of the time. The complexity of reality grows very quickly as you introduce variations to the standard procedure. In the example, the price label on the goods had not been changed and the cash-out operator believed that the item had a special promotional price for that week. Most of us have experienced delays at cash-outs in supermarkets for this and many more small reasons. A process failure before the cash-out is usually the problem but the cash-out operator can often be the victim if the customer's anguish is vented on the luckless operator.

Process charting will also raise a number of issues concerning conventional departmental and functional budgeting procedures. In a typical budget statement for a purchasing department, extracted from the monthly accounts, typically, one finds that the accounts headings are to do with the resources that go into the department. The headings include such things as staff costs, travel, telephone, stationery, premises, computer charges, and so on. However, if we break down the purchasing department into its constituent activities and then ask other departments to do the same then we find that many of the purchasing activities also appear outside of the purchasing department. In Figure 3.8 we see that more than twice as much purchasing activity resides outside the purchasing department. Tracking the purchasing

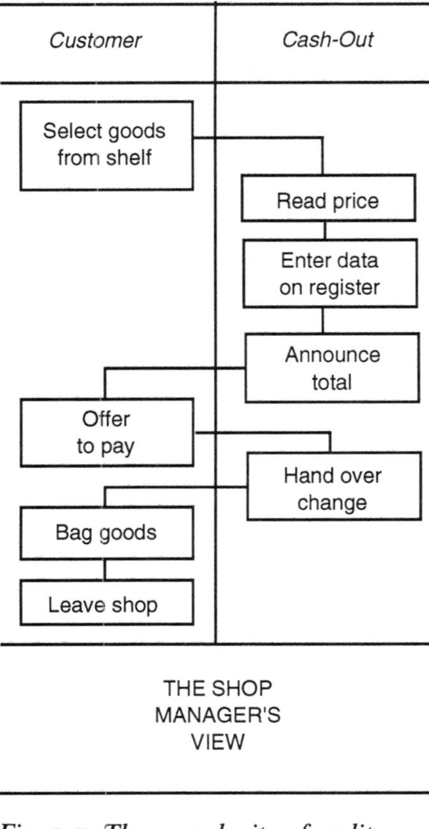

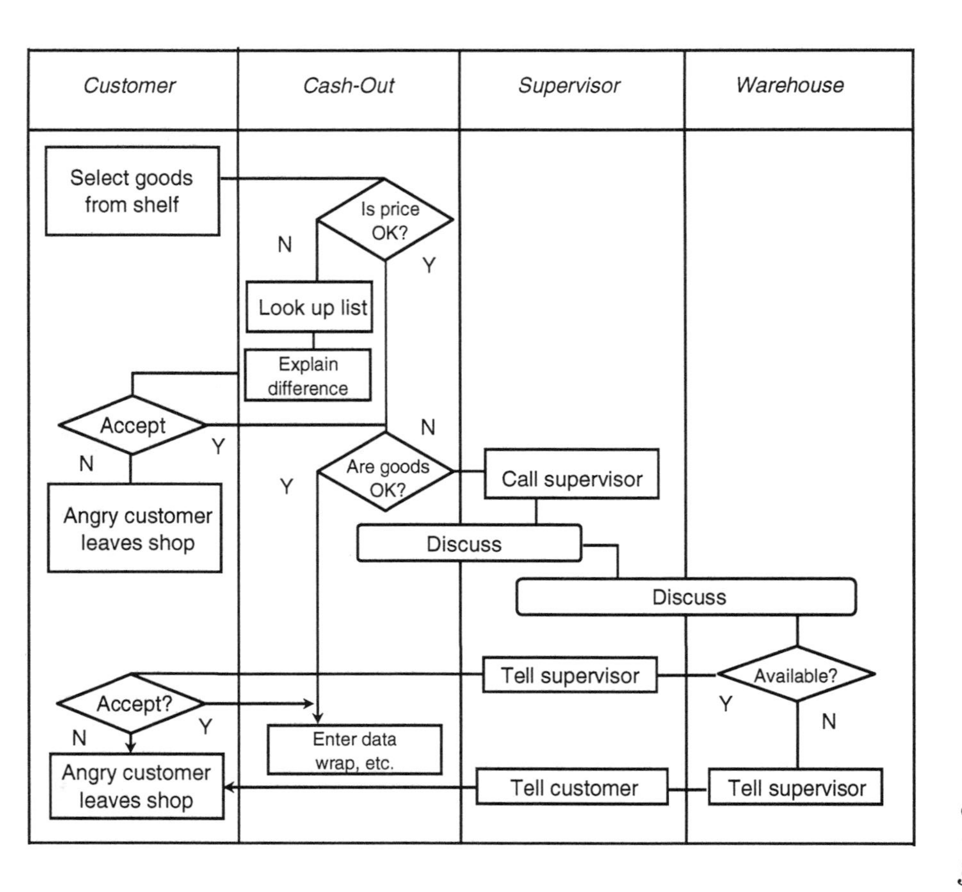

Fig. 3.7 The complexity of reality

Purchasing activities	Purchasing department	Actual costs Other departments	Total company
Assessment of requirements	20	192	212
Requisitioning process	61	203	264
Approvals	12	84	96
Supplier intelligence	6	23	29
Buying negotiations	145	255	400
Progress chasing	116	137	253
Data input	97	–	97
'Noise' (things going wrong]	189	120	309
Direct purchasing activity	646	1014	1660
Management and administration	142	45	187
Communications	69	23	92
Other	51	32	83
Indirect activity	262	100	362
Total activity	908	1114	2022

Fig. 3.8 There is more purchasing activity outside the purchasing department

department's own costs through the monthly accounts will not provide any insights at all into what is driving the purchasing process throughout the company.

One of the core activities for the purchasing department is 'assessing needs'. If the activity were associated with finding suppliers or prices for a part that was on its ninth issue in three months, then we could question what is driving the activity. The design office may have rushed an inferior design into production and the rate of design change then reflects putting the subsequent problems and warranty claims right.

Listening to the voice of the process

When you become practised at using process charts, you begin to hear the 'voice of the process'. In an example from the insurance sector, shown in Figure 3.9 we can hear the process telling us that the external customer is very much part of the process. So much so that the customer begins to regret ever having shown an interest in having a life insurance policy. In head office, there was a general belief that all potential customers understood the quotation, it met their exact needs, they were fit and well, there were no unusual circumstances, all proposals were correctly completed, the final documentation was clear and concise, and all payments were set up correctly every time. After subsequent analysis, only five per cent of new business fell into this ideal category.

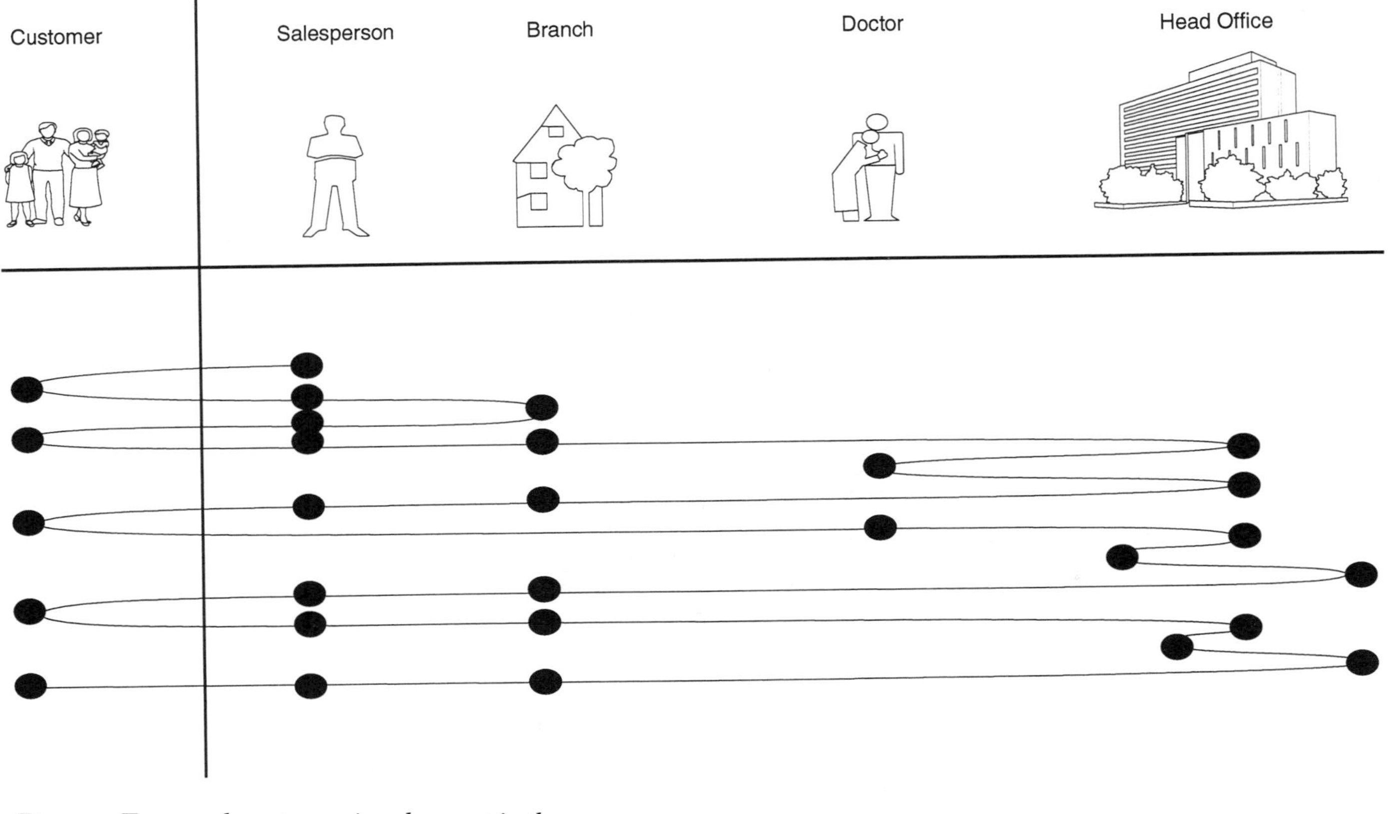

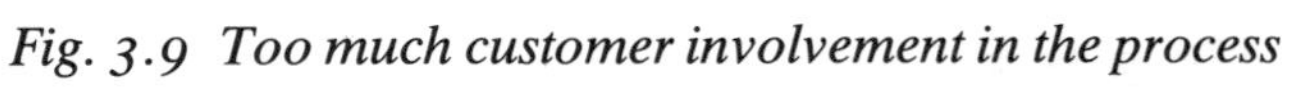

Fig. 3.9 Too much customer involvement in the process

Costing out the activities indicated that checking new business proposals used a disproportionate level of resource. Further analysis allowed the voice of the process to be heard. In Figure 3.10 we see that the most frequent cause of errors was the wrong classification of the population segment based on the customer's job title. By plotting a run chart of frequency over time, it was also found to be a worsening trend. By plotting a series of scatter-diagrams, to look for any relationship between two variables, a correlation was found between length of service of the salesforce and the frequency of errors. The conclusion drawn, confirmed by follow-up, was that the importance of these data to the marketing department had never been indicated to new members of the salesforce. Other than the obvious categories into which job titles fell, little effort was used by the salesforce to ensure a rigorous classification of the customer's population segment.

The conclusions, pointed to a need for better communications and/or training on this particular subject. However, various communications options had different impacts on the problem. An article in the company newspaper had the potential to contact all employees within a short time scale. However the popularity of the paper indicated that only a low uptake of the information could be expected. Also, a newspaper cannot be questioned, should the reader have a query. At the other extreme, a head office course would have a high impact on a small number of people, but it would be a long time before all the salesforce would be covered.

This particular dilemma was resolved at the end of Step 1 when, for other reasons, electronic point-of-sale equipment (EPOS) was made available to the saleforce. Information updates could now be downloaded to each salesperson with tests included to help comprehension. The results of the tests could be uploaded to establish where particular additional help was needed.

Categorising activity data: core, support and diversionary activities (C/S/D)

Among the analyses, one of the most enlightening is the classification of each activity, by the groups themselves, into three categories: core, support, and diversionary. The terms are defined as follows:

Core activities use specific expertise within the group and can be seen to add real value to the business. Core activities are those that provide a necessary service to internal or external customers.

Support activities make it possible for core activities to take place. For example, a salesperson's time spent negotiating with a customer is a core activity. The travelling time to get to the customer is support.

Diversionary activities are caused by a process failure somewhere in the organisation. Such activities include correcting errors, chasing other groups for

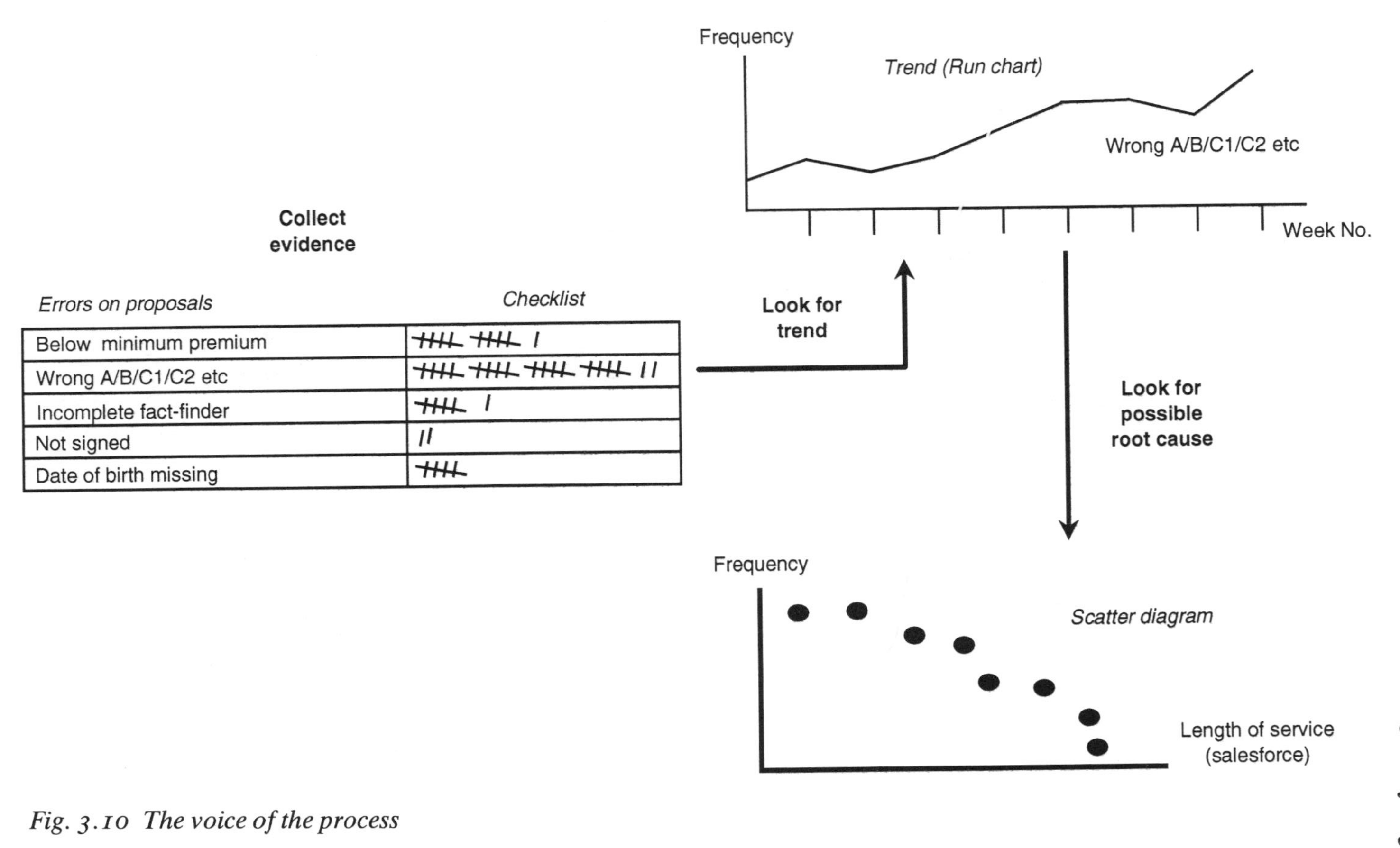

Fig. 3.10 The voice of the process

information, resolving queries, and so forth. Diversionary activities have many causes including, for example:

- inadequate training;
- inadequate tools, procedures and systems;
- poor documentation;
- poor communications;
- poor quality suppliers;
- conflicting functional objectives and performance measures;
- inadequate understanding of customer needs.

Poor efficiency and effectiveness can only be eliminated by isolating the root cause of the problem. Frequently, failures cascade through a number of sections, picking up further diversionary activity, and therefore costs. By identifying the source of failure and the associated diversionary activity costs, wherever they occurred, simple cost/benefit analyses can be undertaken. A key outcome should be to change the mix of core, support and diversionary activity within each area of the business – that is, to place more emphasis on core activity to enhance service quality, and so avoid diversionary activity elsewhere.

The sum of core, support and diversionary activities is a hundred per cent of the activities within the business. In Step 1, all activities will be questioned.

Other types of activity classification

For some companies, the term cost of quality (COQ) has been used to determine the potential benefits of improving the quality of the processes within the business. COQ has three categories:

- the cost of prevention (activities to prevent the occurrence of poor quality, e.g. training);
- the cost of appraisal (activities to find poor quality, e.g. inspection, checking);
- the cost of failures (the results of poor quality, e.g. re-work, scrap).

In general, it has been estimated that the COQ amounts to around thirty per cent of all costs within a business prior to improvements. However, the balance of activity, seventy per cent, is termed the 'basic work'. The focus on just COQ can often lead to the assumption that the basic work does not need to be scrutinised and all attention is focused on those elements falling within the COQ classification.

In the UK, a British Standard has promoted the classification of all activities into being either 'Conforming' or 'Non-conforming'. We would contend that this removes the subtlety of being able to challenge the need to conform to certain procedures. The currently conforming activities may be supporting inadequate processes that are failing to meet customers' needs.

Examples of using C/S/D analysis

Figure 3.11 shows an example of how the salesforce and sales administration department of a manufacturer of office equipment used their time. It was found that only fifteen per cent of the whole department's activity was devoted to customer contact. For a typical salesperson, the core activity of 'selling' occupied only fifty per cent of the time spent with customers. The remainder was spent in dealing with queries and complaints about delivery performance – a diversionary activity. Other activities in head office, such as credit notes, special invoices, keeping statistics on the problems, and a substantial proportion of management and administration was driven by the same problems. Not surprisingly, the saleforce had little time to spend on new calls to win customers.

When the process was investigated each link in the chain can usually point to another link as the cause. If the saleforce got the order details right first time, if

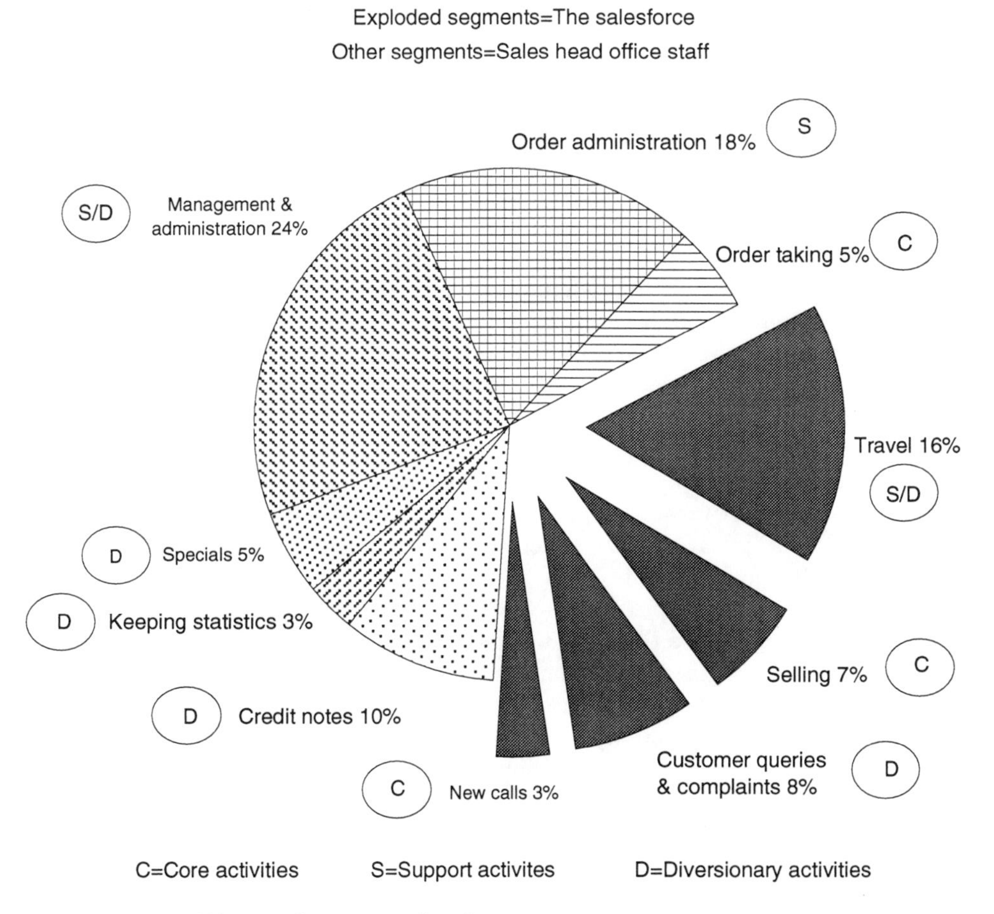

Fig. 3.11 C/S/D analysis in a sales department

manufacturing had organised production schedules to provide adequate inventory, if distribution had kept to the customers' delivery dates, if . . . if . . . if. As one finds everyone attempting to do a good job, then often the cause of process failure can be attributed to each function trying to improve its own, but isolated, functional efficiency. In this case functional effectiveness measures were driving everyone's behaviour. The saleforce was measured on numbers of orders sent in, with month-end panics being the norm. Production ran larger than necessary batches in order to improve machine running-time utilisation. Distributions standard costing variances were reduced by running full loads every time, even if this meant changing customer delivery dates. No end-to-end process measures were in place that related to the processes' ability to satisfy customers at lowest unit cost to the business.

By identifying the root causes of poor delivery performance, a substantial proportion of the salesforce's time was released, allowing it to focus on winning more orders. Time was also saved in sales administration; some taken as a cost saving, while some was re-deployed into dealer support, handling increased volumes and a new task of telesales.

As a general principle, the reported level of diversionary activity in a department is not created by the department itself. After all, it is very rare that we would expect to find people creating diversionary activity for its own sake. We can safely assume that people come to work to want to do a good job and take pride in what they do. Diversionary activity is just as tiring as core activity, but the latter creates job satisfaction, whereas diversionary activity leads to frustration.

It is also often the case that when people are appraised it is on the output of their core activities. High levels of diversionary activity reduce the level or quality of core output but this can be overlooked during the appraisal, particularly if one's manager has not been sufficiently aware or proactive in working across functional boundaries in order to tackle the root causes somewhere else. In the Sales Department example, many of the statistics that were kept listed the frequency that other departments had failed. This evidence was not used to guide process improvements but was guarded carefully to prepare a defence against any charges of incompetence should the business begin to blame the department for declining sales.

Figure 3.12 is an example of the use-of-time analysis, highlighting a problem that had been addressed through misplaced investment. A group of over four hundred product engineers in a manufacturing company spent only twelve per cent of their time on the group's core activities of design and development. The rest was a mixture of activity such as testing and prototyping and a large proportion of management and administration. Much of the management activity was associated with prioritising and re-prioritising the enormous backlog of work that had built up.

It was not surprising, in view of the small proportion of core activity, that the company was recruiting more engineers. In fact, it was the only department that was not affected by the company's recruitment ban. The company had also

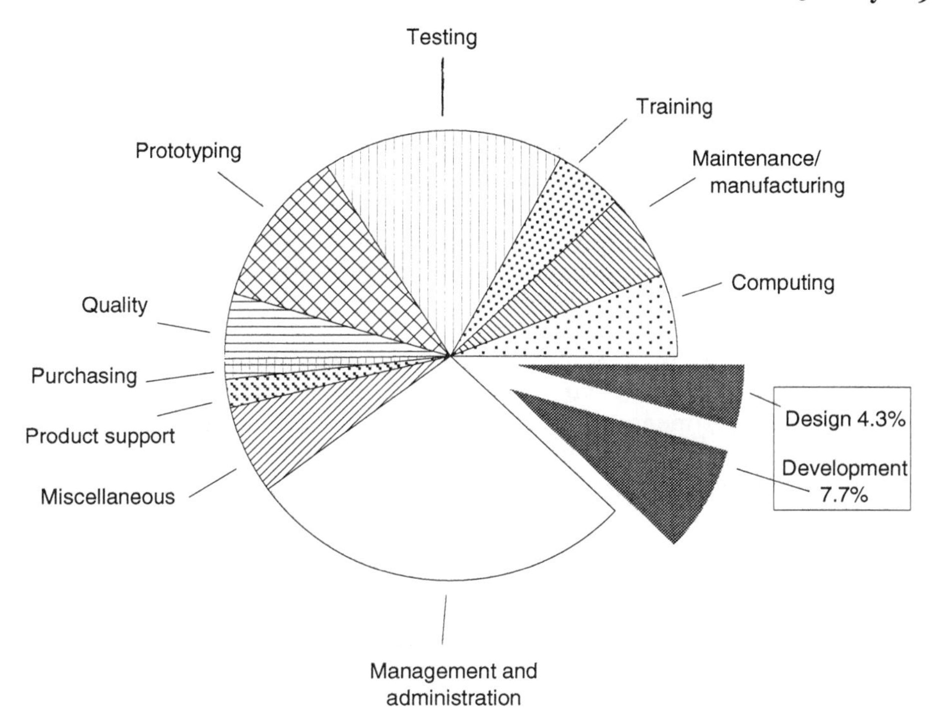

Fig. 3.12 Use of time analysis in a design and development department

invested heavily in computer aided design (CAD) in an attempt to match the lead times for new products being achieved in Japan (half the time) and Germany (two-thirds the time).

The use of time analysis demonstrated that these two actions had a negligible effect on development lead times. New recruits only added an additional twelve per cent of their time and the CAD could only improve the productivity of a small amount of core activity. The answer lay in understanding what was creating the high proportion of non-core activity and then reducing it. This was done in two ways:

- First, by investing in computer simulation to increase the productivity of the prototyping and testing. Many initial design options could be eliminated prior to moving to the prototyping stage.
- Second, by introducing an administration section rather than have expensive engineers perform the same tasks badly, the diversionary time released transferred to core activity for the engineers.

The level of core activity then rose from twelve per cent to forty-one per cent which allowed the CAD to become effective and removed the need for additional recruitment. As the backlog reduced, so did the need to prioritise and more time was released.

The solution, in retrospect, now seems obvious. However, when people work inside routine procedures and standard practices it is difficult to stand above the department and take a more holistic view of what is going on. Activity data, though simple, provides many insights not seen before.

Other uses for activity data

A database of activities can be coded in order to allocate activities to identified processes. This allows the proportions of core, support and diversionary activity to be tracked throughout a process. In Figure 3.13 the proportion of diversionary activity increased as the process flowed through the company. This example shows how a virus, poor order specification, compounded its effect as each department attempted to overcome the difficulties created by the department upstream in the process. The delays created pressure from customers, leading to short-term re-prioritising of manufacture schedules and assembly, and excessive overtime working to meet deadlines.

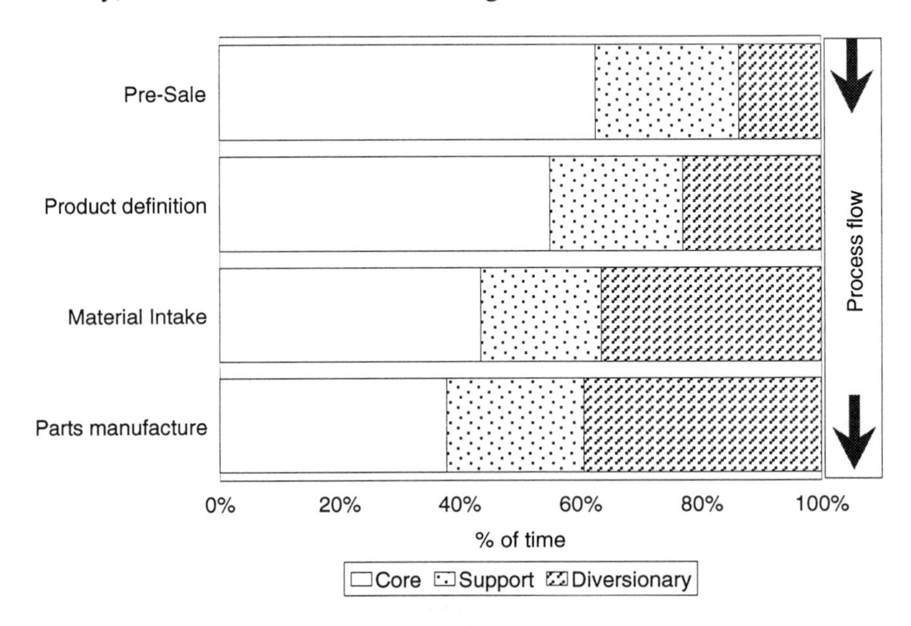

Fig. 3.13 The virus of poor order specification

The final symptom that highlighted the crisis arose when production attempted to overcome the backlog of late deliveries. Working from pre-design sketches, production converted a whole order quantity of raw material into piece parts (aluminium vehicle panels). The customer, on visiting the plant, while delighted that the delivery dates would be met, was curious to know why the specification ordered was not being adhered to. Only a large discount persuaded the customer to take the vehicles as manufactured.

Activity analysis can uncover the true customer profitability. A dry-packaged food manufacturer's overall profits were sixteen per cent of turnover, with variations across the three divisions. However, profits were calculated by using the overhead costs of the staff that worked in the separate divisions. After activity analysis, the staff were asked to allocate a proportion of each activity to the three divisions. This exposed the amount of work that was being done in each division on behalf of the other two. In particular, Retail sales staff were being drawn into supporting the Catering division and Retail manufacturing overheads were being drawn into supporting the Private-Label division. The currently reported small profits in Catering and Private Label then became significantly negative. This evidence prompted the board to take a far more serious look at the whole nature of the business.

The allocation of costs, using activity data, to obtain a better understanding of product and customer profitability is discussed at greater length in chapter 4.

Activity analysis can include the levels of unpaid (and paid) overtime. In particular, unpaid overtime is a measure of organisational and employee stress, and can also represent the amount of goodwill that exists within the business and the tolerance that people develop to the process failures that cause the problems. In the same dry-packaged food manufacturing company as above, unpaid hours were fifteen per cent of normal hours, equivalent to an additional twenty per cent of the salary costs had the overtime been paid for. In Figure 3.14 the relationship between the amount of unpaid overtime and the percentage of core activity for a number of departments shows that the amount of core activity to do the necessary tasks has been created outside of the normal contracted periods. We can also see that the Catering Division was struggling to be a business and that the profitability of the Retail Division masked the real efforts by the salesforce to obtain the business.

It was also the case that in the same company, part of the reward for promotion was ever increasing levels of unpaid overtime. Managers' time went on firefighting and panic actions in a vain attempt to get ahead of the problems. This was a prime example of the headless-chicken syndrome, mentioned in chapter 1.

Attitudes and behaviours

The culture of an organisation can be summarised as the values and beliefs held by most people over a significant period of time, that determine their behaviours and responses to a set of given stimuli. People in an organisation often have difficulty describing its culture. This can be better described by outsiders who have experience of many different types of organisation and are in a position to make more global comparisons. People inside organisations generally see themselves as going to work, rather than contributing to some longer-term goal, and therefore life to them is as they find it with a suspicion that their working life is unlikely to be very different in another organisation doing a similar task.

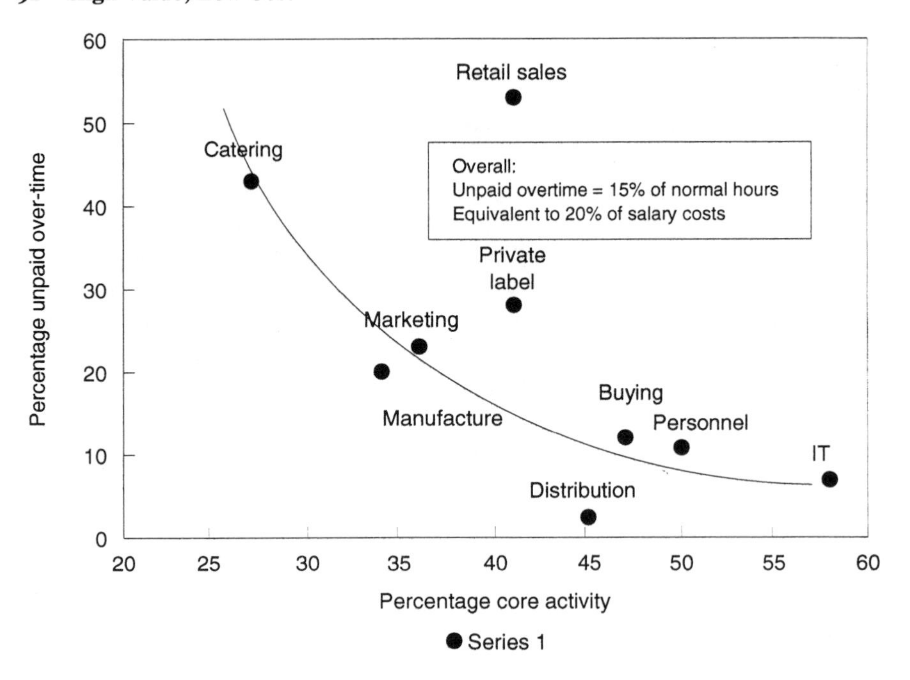

Fig. 3.14 Tolerance to process failures

People link the task they do to their notion of the working environment, rather than think too deeply about what actually motivates them to do the task or to improve it. Their views of 'management' are often expressed in simple terms, such as 'we have good supervisors, we can ask them anything' or 'they never bother to find out my problems, they just chase me for output'. However, a quantified measure of the views that staff and managers hold provides a number of insights into the barriers that will need to be overcome in order to implement change and achieve continuous improvement.

In Figure 3.15, a staff survey highlighted those things perceived to be done well, and those causing a serious issue. The results indicate the level of agreement to a number of questions.

In this example, staff were aware of the external threat, took pride in what they did and understood what the aims of their own functions were. However, they felt that business changes were not communicated well to them, that they had to respond to change without being made fully competent to do so, and that firefighting was the pervading management style. In Business Process Management, surveys should be designed to explore a few key themes as a minimum:

1. The nature of the relationships up and down the structure, at all levels.
2. The nature of the relationships across functions, at all levels.

Poor management behaviours and a poor understanding of cross-functional

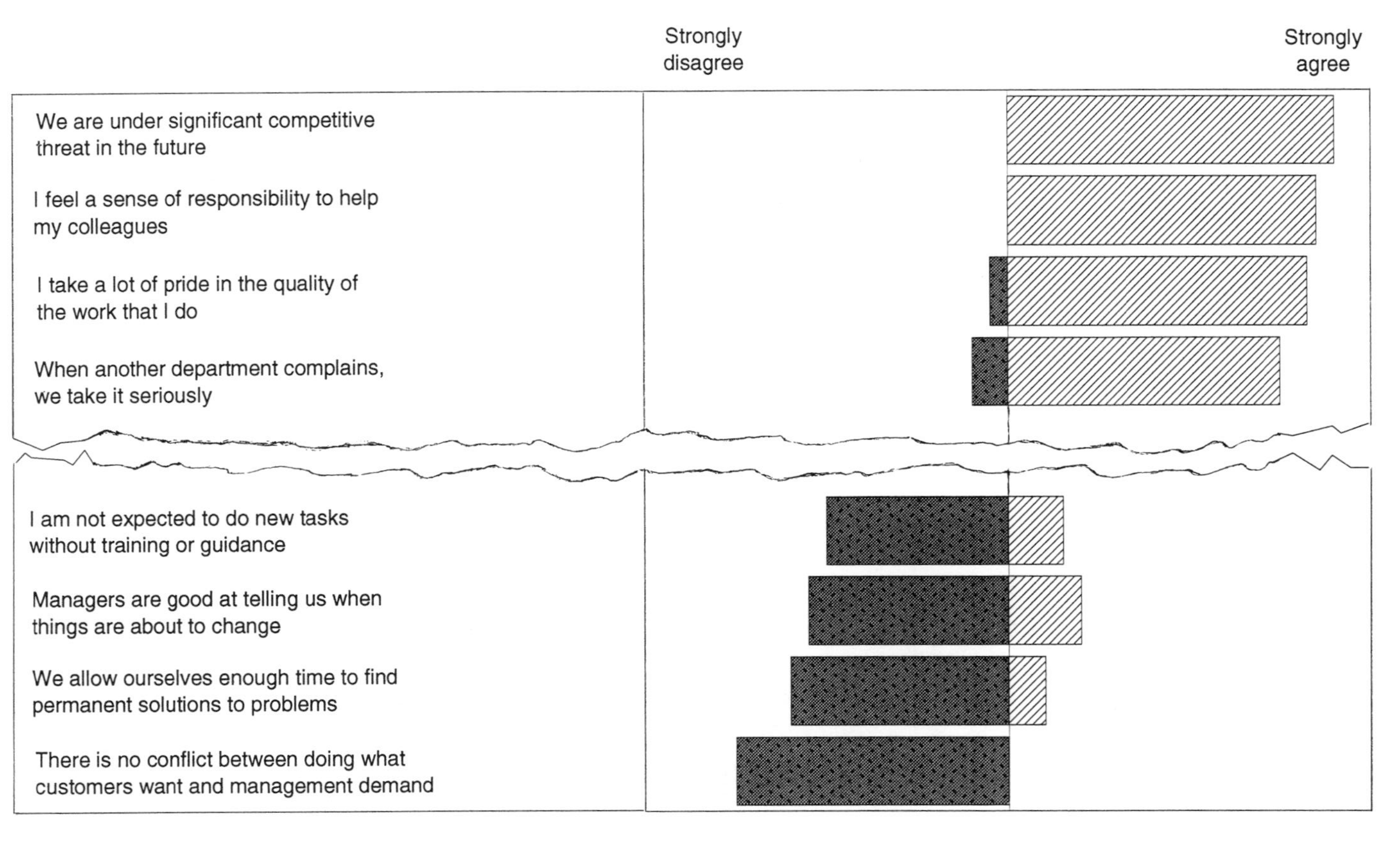

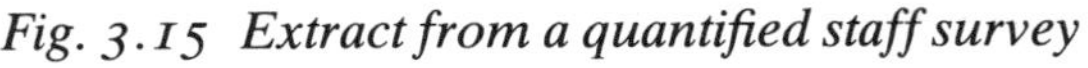

Fig. 3.15 Extract from a quantified staff survey

working in processes are the main barriers to obtaining change. The results can be analysed to compare the views of each function, horizontally and, by management layer, vertically. In the latter case, quite profoundly different perceptions can emerge about the business. Belief in one's self is also a powerful reinforcer of one's style if it is never brought into question. The set of answers shown in Figure 3.16 is a typical response. This type of result can emerge from each level in the structure.

Strongly agree = 5 *Strongly disagree = 0*	*My own view*	*The view I believe my staff hold*
Listening to people who report to managers is important.	5	5
My manager listens to me.	1	5

Fig. 3.16 Belief in one's own style

Collecting data – C1

The purpose of the data is not to expose failures so as to admonish staff. Rather, it is to unlock and remove the veils from people's eyes so they can better appreciate what is actually happening in their own and other groups. The data is the spring board from which to start the process of improvement. As such, the way the data are collected should peel back each layer of the veil, exposing new insights at each stage.

At the start of Step 1, each group leader should involve each member of the group in collecting data. The team of facilitators assist this process and will need to work with departmental specialists or external agencies in order to collect any external data through external customer needs surveys. For the internal data, some guidelines to collect this are shown below.

Organisation chart

The purpose of the chart is to record the current organisational structure of each group, in particular the reporting relationships of all the individuals within it. The summation of all the groups' charts will be the current functional and hierarchical organisation.

Interdepartmental affinities

The purpose is to establish the strength of links between various groups and to identify opportunities for merging or de-merging any groups, where this could significantly improve the processes in which they work. Of particular interest should be the degree to which information is shared, the frequency of communication, and the level of dependency on services provided and received.

Process charts

The purpose of the process chart is to establish the relationship between main activities and decision points within the group and the process connections to other groups. The facilitator should assist this task by asking each member of the group to build up the overall picture. The facilitator can overcome the problem where members of the group would tend to chart only how the process should work rather than how it actually works in practice. Whenever parts of the chart link to other groups then members of both groups and their facilitators can work together to ensure there are no loose process ends.

The 'cast of characters' across the top of the chart show who is involved in the process. The vertical axis can be used to signify elapsed time. An improved process could involve less people, using better methods over a shorter elapsed time. In Figure 3.17 the major frustration felt by the National Account Managers (NAMs) at the company's inability to negotiate quickly with customers was exposed by charting the process. The new process, shown in Figure 3.18, agreed to with enthusiasm by all parties, began to give the company real competitive advantage.

In another example, three divisions of a company charted their processes. Their casts of characters were very similar, but the functions were in the separate divisions. Each chart included a column for the customer. As more detail was added quite complex loops involved the customer and the various queries they raised concerning deliveries and invoicing. The same customer featured on the charts of each division. The customer was attracted by the overall proposition of obtaining products A, B and C, but then fell victim to the three salesforces with separate bonus structures and motivations, and the three different invoicing processes. Customers who had difficulty recognising the three divisions would pay the well-known main company name, creating many queries and mistakes as the business tried to unravel what had happened.

In a research establishment, the Engineering Standards department charted their processes. All seemed quite innocent until the frequency of contact with other departments was measured and the activity data were established. It transpired that the Standards department was busy re-drawing a bought-in service microfilm index of 25,000 proprietary parts and giving each part a new internal number which was entered into the internal stores catalogue. The manual catalogue was so large that users invariably telephoned the Standards department to find out the part number to put on the requisition. At the other end of the process, Goods Inwards had to contact the Standards department each time parts arrived in order to find the internal number that allowed delivery to be noted in the computer system.

The records showed that over two years, only 350 parts had any movement on them. However, the catalogue did include six different sizes and colours of

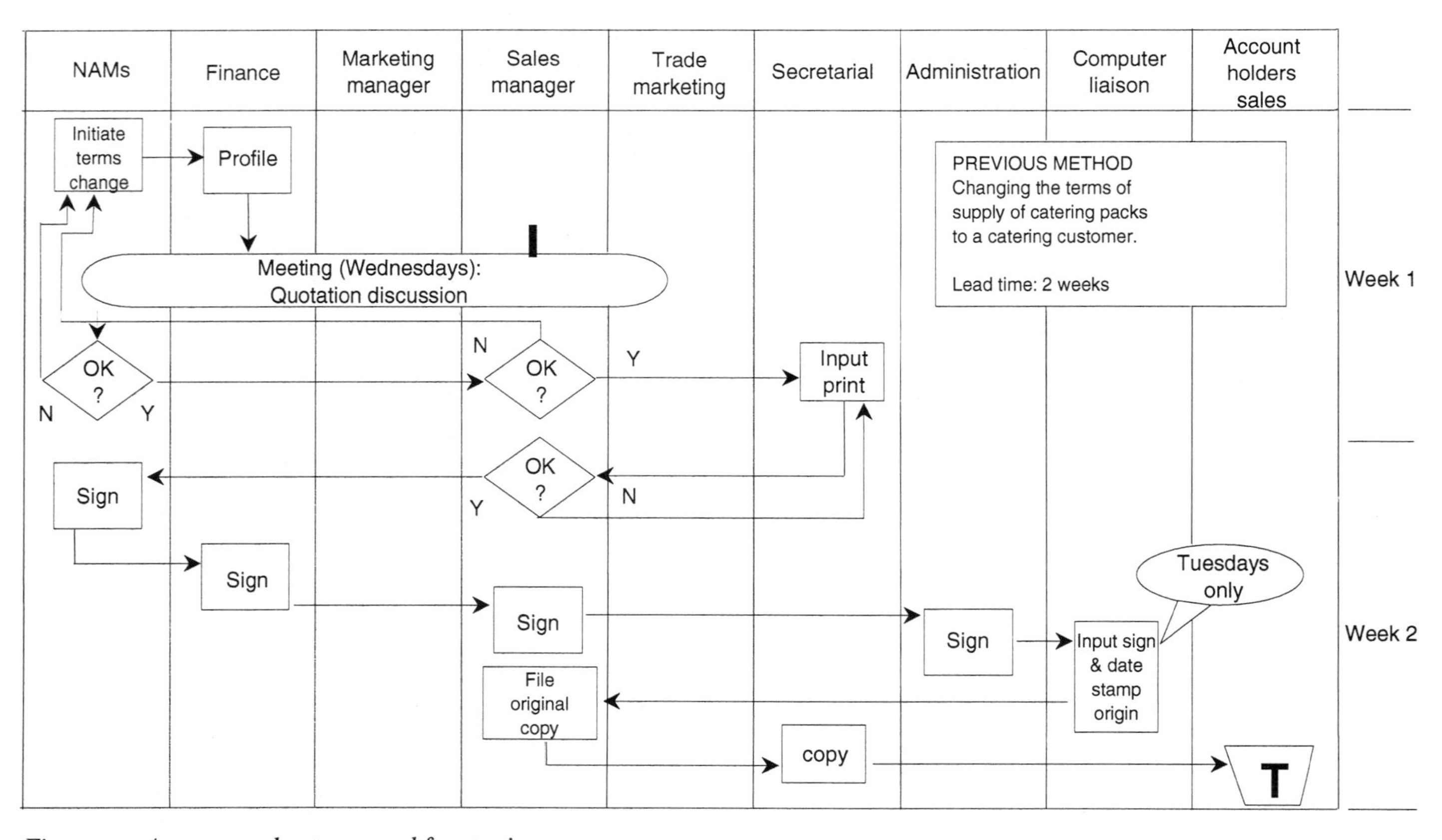

Fig. 3.17 A process chart exposed frustrations

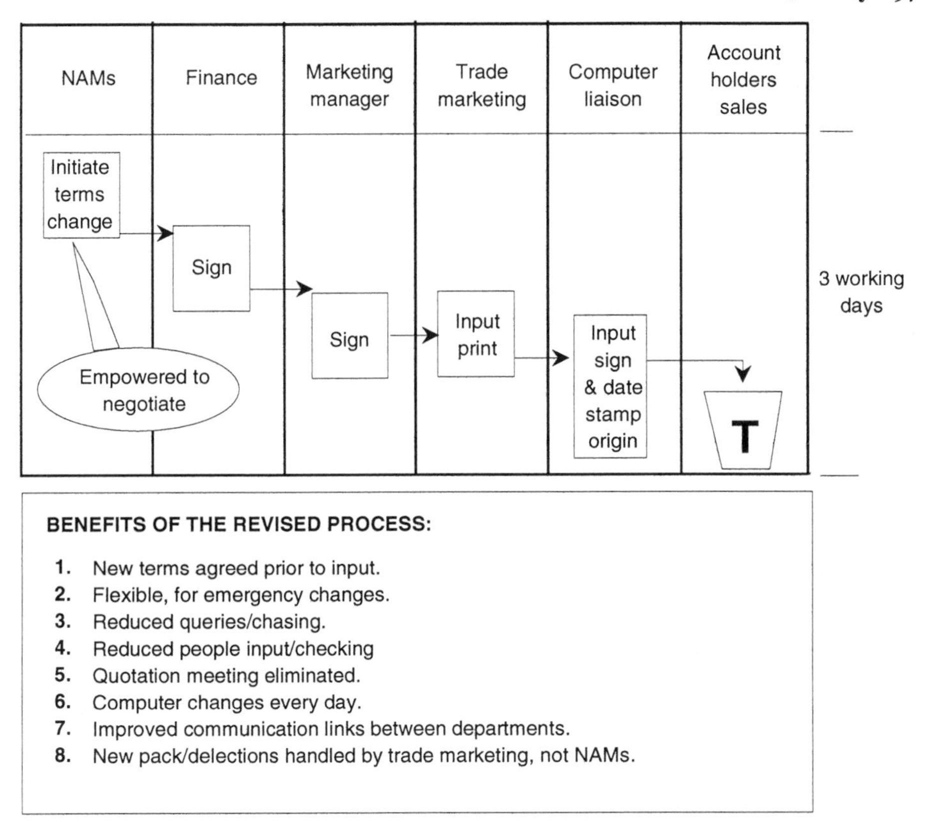

Fig. 3.18 The new process gives competitive edge

paperclip. Quite a lot of management and administrative time was devoted to travelling around the four other research divisions to attempt to reconcile the different part numbers for the same item that were busy being created but from different source material. All the similar Standards departments were busy creating their own industry, independent from meeting users' real needs.

This example shows that just drawing a process chart is insufficient to understand fully what is going on. Improving the process was not the real issue. In this case, giving users computer access to the current catalogue, with key word searching, coupled with screen enquiries to the Standards department for non-stock items, provided the main change. The activity previously handled by thirteen staff was then handled by two, allowing the rest to be re-allocated to real added value activities.

Activities and C/S/D

The purpose of collecting activity data is to assemble a picture of the use of time within each group and to categorise each sub-activity according to whether it is core, support or diversionary. A meeting of the members of the group provides an opportunity for everyone to agree the process chart and then to list the main activities undertaken within the group. Great accuracy is not required, this is not the basis for measuring an individual's productivity against a standard. Such an action defeats the whole object of establishing the links in processes. Even at this stage, the proportions of time on each activity could be a surprise to the group.

Reading these data with the process chart can highlight where the chart has understated the complexity of reality and the process chart can be updated as more complexity is uncovered. To help prompt ideas for changing the processes, the group needs to ensure that it has included those activities which represent work caused by process failures elsewhere in the system, and identified these separately (e.g. correcting data, answering queries, progress chasing, and so on). This will be useful later on when the amount of resource required to address such problems is discussed.

At this point, the group can reconvene to take a hard look at the inputs it receives from other departments. Where the process starts somewhere else, the group looks at the inputs and establishes how important the inputs are to their own work. The group then takes a view as to the performance of the internal supplier, taking into account timeliness, accuracy, frequency, and so on. This provides a measure of the 'Quality' as perceived by the internal customer.

The vital next stage is to categorise the activities of the group into core, support and diversionary. The service-received analysis is now a key input to help the debate. A number of points are important at this stage:

1. It is important to remember that the classification of an activity as diversionary is not a comment on an individual's productivity or performance – the purpose of classifying activities is to reveal where the business processes are falling down. Nobody comes to work with the intention of causing diversionary activity.
2. Many of the activities will be purely core, support or diversionary, but some will contain elements of two or even all three classifications.
3. There are no 'right' answers, the group enters the classifications that they believe are correct from their point of view. All the activities of a customer complaints section could be classified as diversionary, since if other failures within the company's procedures had not occurred then complaints from customers would not exist. However, while complaints still exist, the group would still find that its own time would fall into the three categories.
4. The group will need to be aware of activities that are typically looking for consistent errors they always seem to receive from another section. The first view could be that such checking is core, as to stop checking is likely to lead to errors in their own outputs. Such checking is diversionary but cannot be stopped until the certainty of error-free inputs is at a consistent high level. This will come with a permanent improvement to the process.

5. Core activities are important; in contrast, diversionary activities always seem to be urgent. 'Urgent' takes precedence over 'important' and often the urgent activity needs the most skilled member of the group to sort out the problem.

As Step 1 advances and such processes become clearer, it is essential to see how activities within each group are linked to particular processes. During Step 1 we are considering the current processes. At this stage, the current processes within the organisation can be determined with reference to the overall set of process charts from all the groups. By coding the processes and adding this information to the activity data, the overall core, support and diversionary activities by process can be determined.

Where activities are directly linked to sales volumes then it would be expected that the resources would also need to change. In other instances, the activities are not so clearly linked. A design office, for example, may be driven by the need to overcome previous design problems highlighted by rising warranty claims. A purchasing department may find a stable workload change to an ongoing decreased workload as a policy of having a smaller number of assured suppliers is implemented. Monthly work-in-progress valuations could be driven by a number of factors – range of part numbers, number of parts, location of parts, frequency of valuation. The demand for savings accounts at a bank may be driven by macro-economic trends over a period of time or quickly after a national budget is announced.

In attempting to set a budget for any group it is first necessary for that group to understand fully what is likely to change the level of resources it needs. In many companies, this relationship is poorly understood, leading to frustration at budget time. Often, the budget process is one of multiple iterations as attempts are made to create an overall total that the board have previously decided.

Further data can begin the process of unravelling the cause-and-effect between departments and link this to the database of activities created by each group. On the list of activities the group discusses and then notes what they believe appears to 'drive' the volume of activity; that is to say, what causes the amount of that activity to vary. Very often, the real driver of an activity may be a long way from the group and will only be seen once a number of groups are found to be in one continuous process.

Customers provide a business will all its costs before providing revenue. The data can be used to understand the activities that are driven by different types of customer and to what extent they vary depending on the nature of the relationship with the customers. By knowing the revenue by customer type, customer profitability can be calculated. Part of Step 1 can then include 'customer engineering' in order to turn low or unprofitable business back to acceptable levels by eliminating it, changing prices, or working with the customers to overcome process viruses that are cross-infecting the two businesses.

Root causes

At this stage, the group will have a clearer idea of where to start collecting additional data to look for root causes of problems and to establish priorities to resolve them. The reasons the data are incorrect can be logged and the frequency by cause found. The frequency over periods of time will show the trend. A worsening trend of a frequent problem would attract a high priority when looking for ways to resolve the problem.

Each type of problem will have different implications both to the customer and the supplier. Some low frequency problems may have serious implications. Setting priorities for action needs to take account of frequency, implications and trends.

Staff surveys

In a culture that is enjoying an environment of management help, with properly motivated people, one can argue that a staff survey would have a hundred per cent return rate but would not bring anything new to the surface. Full achievement of this condition is rare. At the other end of the scale, a fear driven, results oriented environment, with people controlled in poor processes, in itself, would ensure that any survey would not bring out the real issues, even from the few people who were brave enough to complete the questionnaire.

Designing a questionnaire and creating an atmosphere that ensures that a high and honest return is achieved are the prerequisites needed to enable meaningful conclusions to be drawn. A high and honest return can be achieved where people believe that the outcome will be acted upon. Too many surveys without follow-up action lead quickly to a declining response rate. An atmosphere of lack of trust in management will also reduce response rates as people remain fearful of some act of retribution for making any sort of negative comment. Using a third party to undertake the survey and making all the summarised results public is one way of overcoming the major barrier.

An initial design for a questionnaire will need to be tested on a randomly selected number of people from within the organisation, the mix in the sample being representative of the numbers of people in different functions and grades. In one company, staff in field-operations reported a lack of understanding of the phrase 'internal customer'. Field staff were so used to the notion of 'customer' being the public that they could not grasp the notion of an internal customer other than in terms of the company's own staff buying their own products.

If the need to undertake BPM is apparent to the organisation then it is likely that the survey will extract even more dark elements. Issues to do with both cross-functional process failures and poor management styles will be exposed. Neither issue can be ignored as staff could be seen to be the victims of poor processes and management, but managers themselves are the victims of being products of the overall business culture. BPM should not be seen as a process to create victims or to lay the blame at individuals' feet. Changing the capability of the business therefore has to include improving management's capability though learning a better way. Unlike a process, management's learning usually has to follow a period of un-learning in order to break with the styles of the past.

CHANGING THE CAPABILITY – C2

An organisation needs to be viewed as a set of multi-functional processes that supply products and services to customers. All the groups need to work as one company, meeting the needs of the internal customers in a way that aligns with meeting the needs of external customers. The processes that ensure this do not respect any internal divisions – every group has a vital part to play.

Generating proposals to improve processes

The next stage is to look at the many ways that all the processes can be improved. As all the people in the company work in these processes, they will be aware of the things that go well and those that do not. The next stage is thus everyone's opportunity to propose improvements. Employees (the victims of poor processes) have the knowledge of why things fail and can therefore propose many ideas for improvement. However, any proposals for change inevitably affect other groups, either as receivers of a service provided by a section or as service providers to the section. By exploring the impacts of proposals for change, cross-functional awareness increases, internal barriers are removed and the cause and effect of change is understood.

Having collected a quantified database covering all the activities which are carried out within each group, the groups will now be able to quantify the costs and benefits of ideas for change. The database is the starting point from which to develop alternative options for the future. The use of time analysis, in particular, provides a focus for the groups to concentrate on obtaining the most cost-effective balance of activity. Diversionary activity cannot be eliminated entirely, but must be minimised by identifying, then fixing, the root cause. Support activity must be performed, but can its efficiency be increased by improved systems or procedures? The same question can be asked of core activities, but it is more important to establish the benefits of enhancing core activity or the risks of decreasing it, both to internal and external customers.

Preparing for brainstorming

The main purpose of a brainstorming session is to examine ways in which the business processes may be changed by:

- improving efficiency through changes to procedures, methods or systems;

- improving the service provided by other groups so that diversionary activity is reduced;
- reducing the service levels provided to others, particularly where the risk is minimal;
- developing and training of staff to improve speed and reliability;
- increasing the service levels provided to others, where there is a clear benefit;
- undertaking new tasks where there is a demand from an internal or external customer, and clear benefit to the organisation.

In order to maximise the input from everyone's experience, the leader needs to involve as many people as possible in the idea-generation process. Each member of the group will have contributed parts of the group's database. As early as possible before holding a brainstorming session, each member of the group needs to study the full set of data from the whole group. Prior to the meeting, members of the group can begin to ask themselves a series of questions that will prime their thinking and increase their contribution to the brainstorming session. The aim of the brainstorming process is to generate ideas which will solve business problems affecting the group, its own effect on other groups, and the effect other groups have on itself.

If the work on positioning the business in the long term is underway or complete, then the conclusions from this research may have been deployed throughout the business. This will provide a higher degree of focus to the groups when they are generating ideas for improvements. However, if the positioning work indicates that the level of change required in the longer term will need to be radical and innovatory, then bringing this to the attention of the groups at this stage may only serve to confuse people as they will have difficulty relating to the bigger business themes. At this stage, the groups need to take account of any positioning conclusions, should they exist at this stage, in order to avoid suggesting process changes that could make more radical change difficult during Step 2.

Some of the ideas generated during Step 1 will be radical and innovatory but they will be out of context if the work on positioning is not complete. Such ideas will need to be brought back into Step 2 for inclusion, or otherwise, at that time.

Holding a brainstorming session

The facilitator from the support team will need to have the skills to facilitate brainstorming sessions effectively. Although the facilitator can run the meeting, it is important that ideas are generated by the group, and not the facilitator.

No limitation on the scope of ideas should be permitted to ensure that all opportunities are explored and to prevent instances of management intimidation that can arise on some occasions. Although brainstorming is a simple but powerful technique used in many companies, there are still many instances of members of staff leaving such sessions and commenting that they have worked for the organisation for over ten years and nobody has ever asked for their opinion on the work they and others do. As well as tapping into a rich vein of ideas, brainstorming provides a clear signal from the business that it respects its employees' views. Managers discrediting all ideas emerging from their staff at brainstorming sessions not only block the flow of ideas but demonstrate how they will become a barrier to improving the business. Such a stance would be unwise to hold throughout the duration of Step 1. These cultural change (or barrier) signals should not be underestimated in terms of the leverage they have on the outcome.

One approach to brainstorming is to invite a free-for-all generation of ideas from the group. This has the disadvantage of allowing the proactive members to command most of the time allowed for the session and leaves the more timid members resentful that they were not given time to express their concerns and ideas. Leaving this impression with some people is to risk raising their scepticism about the whole process of achieving improvements.

The facilitator needs to bring structure to the session without stifling ideas. A large version of the process chart on the wall will prompt the group to cover all aspects of all processes, and the links to other groups. The set of the database, and other analyses, in front of each member provides more prompts than would arise from using their memory alone. The development of ideas can also be structured by introducing a simple classification. This can take the form of:

- improvements in methods and procedures in the group's own working environment;
- system changes to modify or introduce computerised procedures;
- conditional improvements – those parts of the process that need to be changed upstream from the group in order to improve the 'Quality' of the inputs they receive, the reduction in diversionary activities in the group being conditional on changes made somewhere else in the process;
- reduced service from the group – these proposals reduce the 'Quality' of the group's output in some way. Although this appears at first sight to be in contradiction with Step 1, it will often be the case that the demands of some internal receivers of a service may be to serve a parochial need that may not be supporting the needs of the business or the external customer.

The list so far would generate spare capacity within the group. Ideas to re-use this capacity also need to be uncovered.

- enhanced service – these proposals improve the 'Quality' of the outputs from the group and should lever improvements downstream from the group by reducing another group's diversionary activities. The enhanced service eventually, or directly from the group, should impact the external customer. This contributes to creating the changed positioning from P1 to P2;
- new tasks that the group proposes to generate a benefit to the business

greater than the resources to undertake the tasks – these also contribute to creating the changed positioning from P1 to P2.

Although many analyses are prepared by the group prior to brainstorming, there will always be a number of problems raised at the brainstorming. A simple technique to use is the cause and effect diagram. Figure 3.19 shows the conventional approach. The main headings, namely Materials, People, Methods and Equipment, are those factors that contribute to a particular desired effect, such as 'improve data input accuracy'.

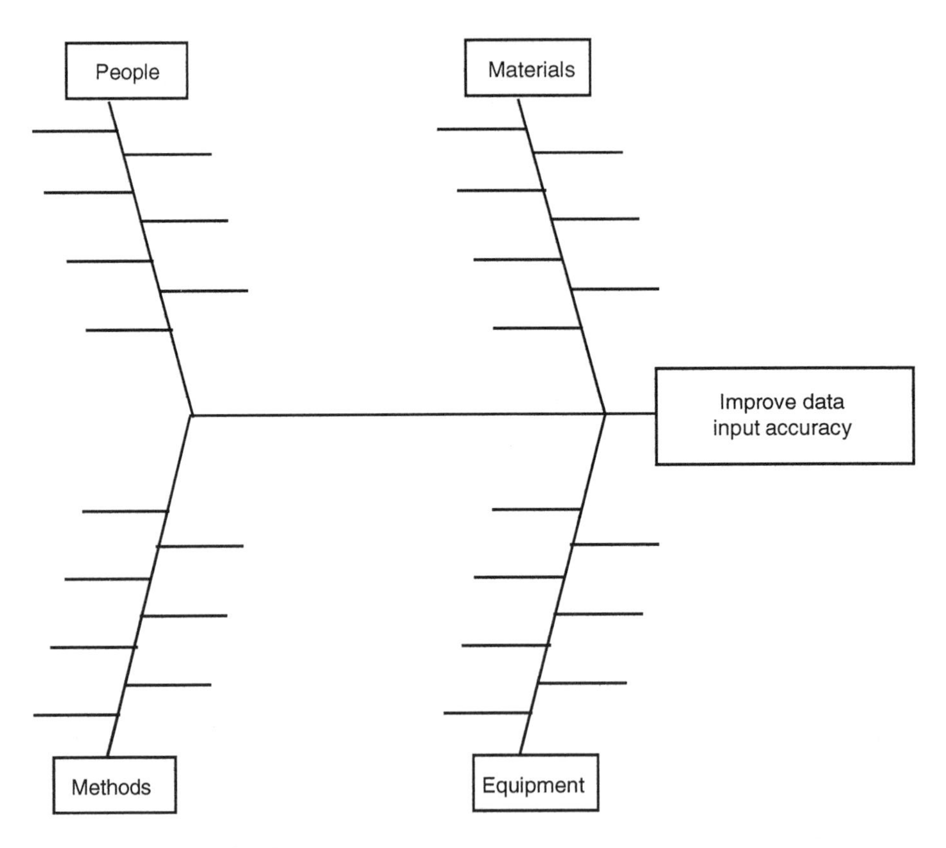

Fig. 3.19 A standard cause and effect diagram

However, practice has shown that to create a 'reverse' cause and effect diagram generates more ideas. For example, if the problem is to 'improve customer invoicing errors', then the diagram can be drawn to find ideas to create just the opposite effect; in this case, 'to ensure all invoices contain errors'. Experience has shown that people remember far more instances of things going wrong, when they are asked to create problems, than when they are asked to resolve a problem.

Figure 3.20 shows how this approach was used to generate as many ideas as possible to ensure that a 'perfectly disastrous meeting' would be held. Most of the causes came from various individuals' personal experiences. The opposite of all the root causes then becomes the proposals to avoid a disastrous meeting. The proposals also create a process for setting up meetings and a checklist to ensure that each step has been completed.

If the group is uncertain as to the severity of the potential root causes, then more data on the frequency of the events, and the implications when they occur, will need to be gathered after the brainstorming session.

Another technique to use during brainstorming is force-field analysis. In this approach, the group explores the forces that are acting for and against an action taking place. For example, taking one of the causes for concern, such as 'data input accuracy', the group lists the negative forces (poor design of the input form, inadequate lighting, noise level, size of backlog) and then lists the opposing positive forces (error trapping built into software, job rotation to alleviate strain). The brainstorming session then extends both lists to provide ideas to improve the process.

Reviewing the proposals with internal customers and suppliers

Many of the proposals for change will inevitably affect other groups, either as receivers of a service or as service providers. A group leader, along with other specialists from the group, will need to hold a review meeting with representatives from their service-receiving and service-providing groups to discuss the draft proposals that may affect them. A review meeting represents a business process from one group's perspective. Other review meetings will also consider the same process, but from other groups' perspectives. Overall, every process is thus considered in a multi-functional manner.

The review meetings provide a vehicle to expose the complexity of reality and increase the awareness of the participants of the relationships within a process and the cause and effect of changes within it. In particular, the review meetings need to consider proposals that:

- require service providers to improve their service;
- reduce or enhance the service levels provided to external customers or internal service receivers.

Preparing for a review meeting

To maximise the degree of understanding and to provide a structure for the review meeting, the facilitator from the support team and a group leader prepare the agenda for the group based on the ideas from the brainstorming. Although the group can quantify the impact on its own diversionary activities, it will be necessary to address the cause of the problem somewhere else during implementation. In many cases the group will be aware of the symptoms it

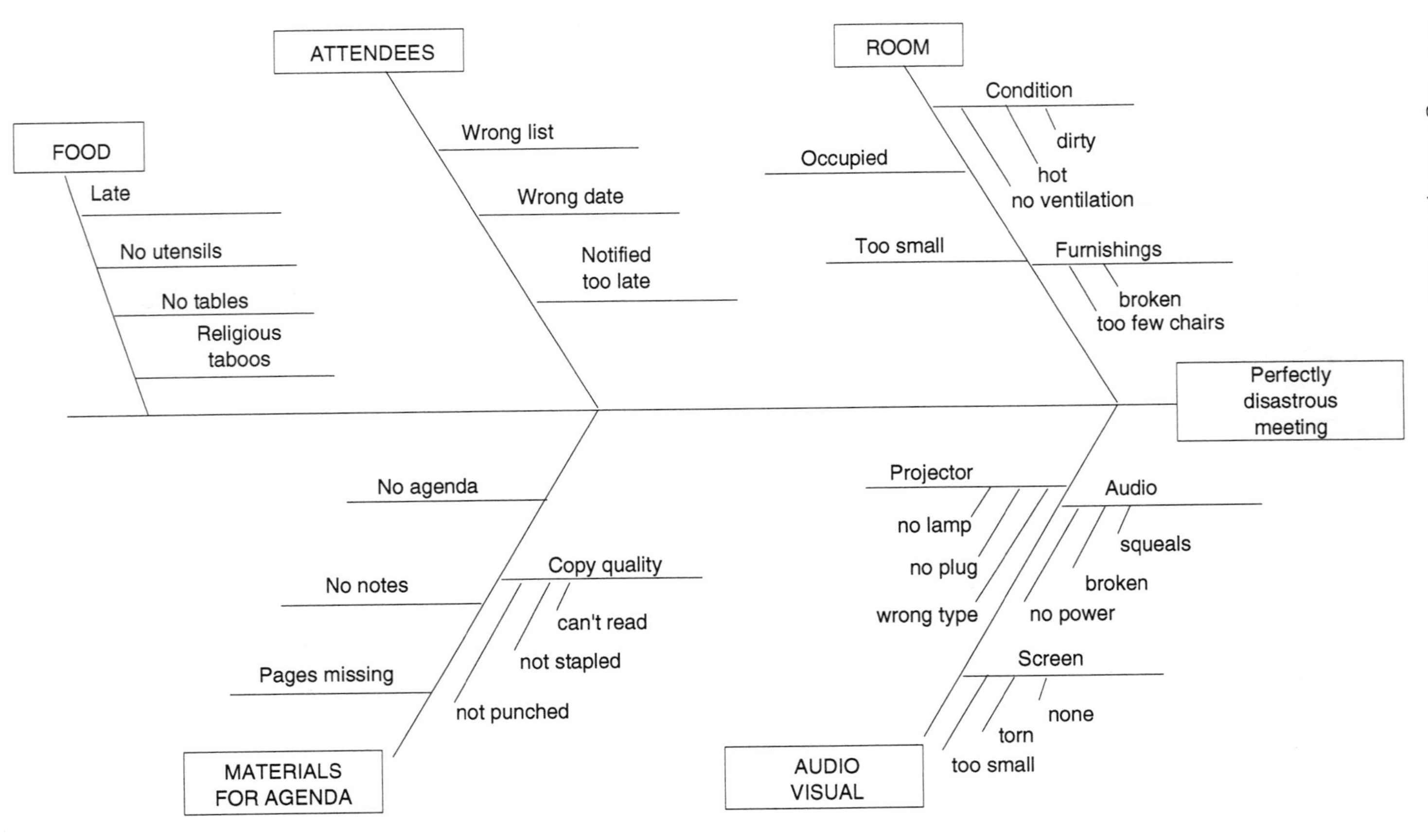

Fig. 3.20 A reverse cause and effect diagram

suffers from a process failure. However, the root cause could be much further away from that group in the overall process. The root cause may only become apparent at the review meeting, when the whole process is visible.

Some proposals to reduce service will not incur risks to the business or reduce external customer service. Rather, they will be the elimination of outdated procedures that have been perpetuated for many years without question. Other proposals to reduce service would impact on external customers. The proposals are genuine options and may have to be taken seriously if short-term cost reduction is a necessity. Reductions in service will tend to generate risks to the business. There is always a balance between the nature of the reduction, to maximise the saving, and the level of acceptable risk to the business. This level will be very judgemental and will create much debate.

The reduced service or prioritisation will create some risk in business terms (lost sales, incorrect deliveries, etc) or as an effect on another group (late data, infrequent updates, etc). The riskier proposals will be of particular interest to the group's internal customers who will be those who would experience the risk. Some proposals, however, may be seen as creating a risk at a local level, but no major risk would occur at a company level. Some proposals would not be risky if another group changed its service (e.g. Quality Assured suppliers would reduce the risk of reducing Goods Inwards inspection).

The enhancements need to focus on core activities, to give the best return to the business.

Review meetings

The review meeting agenda is created from the list of brainstormed ideas. After the agendas have been prepared, these are discussed with the group and its internal customers and internal suppliers at a review meeting. To provide impartiality, the meeting should be chaired by a facilitator from the support team. The participants at the review meeting are determined by the group referring to the database, particularly affinities, input performance, and the process charts.

The purpose of the meeting is to help the group decide what the level of service should be, in the light of the risks and benefits of change, the reality of achieving implementation of the conditional proposals, and the affordable level of costs within the business. The participants will all be part of the chain of activities that make a process. As the key business processes are multifunctional and run horizontally through the company, it is vital that review meetings should expose all the cross-functional issues.

Wherever possible, the review meetings should be encouraged to make decisions when all the key elements of the process chain are represented and all the implications of change have been debated. In some cases, the participants as a group will assume a level of empowerment beyond that normally assumed by any of the participants acting individually. The degree to which the participants do make decisions will also provide a clear signal as to the level of empowerment that is allowed or assumed to be allowed. However, some proposals, though agreed to by the participants, may still require a higher level authority or decision-making procedure before being finalised for implementation.

Management brainstorming

The approach described so far has a strong emphasis on being applied bottom-up in the organisation. In other words, starting with those people doing the everyday work inside processes. In most organisations we can assume the following:

1. Managers will want to contribute their own ideas for improvement and may sense the same frustrations as their staff in not having been listened to in the past.
2. A bottom-up process will leave managers feeling exposed when their own staff raise many examples of process failures that the managers, in good faith, believed did not exist.
3. The process will expose managers who have believed they have been able to hide problems.
4. Managers will have a better overview of cross-functional processes so their views will add value and lead to improvements.
5. An employee attitude survey will expose all the management styles and the poor managers will need every opportunity to be given help and knowledge, even when their expectation is one of being pushed into the abyss.

The outputs of data collection and analysis, prepared to highlight the key business processes, form the inputs to management brainstorming sessions. The issues arising from the groups' brainstorming sessions, particularly the conditional proposals that impact along a process, provide a key focus for the managers' sessions.

The support team facilitators should be well positioned to facilitate the managers' brainstorming sessions as they will have been immersed in cross-functional issues from the start of Step 1. As well as business process issues, the sessions should also include the summary analysis from the staff and management surveys.

A brainstorming session can start the teaming process, if it does not exist already, and provides a relatively neutral starting point to begin to explore the role of the manager after completion of Step 1. Managers thus take two routes on the BPM journey during Step 1:

1. During a series of workshop sessions, managers come to terms with the diversity, history and consequences of the styles and behaviours they have currently and start the learning process to discover other roles and behavioural options. This can be a painful experience for some as previously accepted and, from their perspective, successful behaviours are challenged.

2. The emerging proposals from the groups will need to be consolidated with their own ideas and then implemented, and in some cases authorised for implementation or otherwise. This will require that managers begin to apply new behaviours and begin the new role they need to adopt in order to make continuous improvement part of everyday business-as-usual.

Agreement to change

During Step 1, the business will remain a functional organisation. Changes within the function that have little impact outside the function will have been debated and decisions made during the review meetings. Some functional issues requiring higher authority will require consideration within the hierarchy of the function. A functional decision meeting, with the group leaders and functional managers as participants, is used to create the preliminary functional plan for implementation of its own changes to methods and systems. The database of actions, created at the review meetings provides the basis from which to track implementation.

At this stage in Step 1, all conditional proposals will have been discussed and some decisions made during the review meetings. The support team will need to sort out all such proposals and consolidate all the requirements to each function from every other group. A functional decision meeting also provides the vehicle to understand the implications on the function of the demands placed on it by all its internal customers. There will also be a number of overall functional implications due to:

- the service-level change proposals from and to the whole function;
- to and from all other groups;
- and the changes so far agreed at review meetings.

At this point, the conflict between functional needs and cross-functional process needs is brought into sharp focus.

As managers will have already participated in their own brainstorming and orientation workshops, the focus of their approach should now be changing towards processes while still managing inside their functions. The output from the functional decision meetings should now be brought together at a forum of managers that represents each key business process. The functional implications of conditional proposals and service-level proposals now have a proper context as the implications of the process on the external customers, changes to provide P2, are embraced in a discussion of the mechanism to change the capability C2. The overall benefits of P2 to the business can now be reconciled with the cost of the capability to be implemented as C2.

Consolidating the plan to achieve P2/C2

To consolidate the overall plan and to ensure that each key process is contributing to achieving an overall P2 position for the organisation, it will be necessary to bring the senior management team together at a workshop session. The plans and decisions from each lower level meeting will have given a message relating to the emphasis to be given to particular options proposed for each key process. Often, not all the enhancements offered are seen as having the same short-term priority when viewed from an overall strategic business perspective. Also, some of the proposals may have the effect of changing the activity volume in certain groups which will impact on the resource levels so far determined for a particular group in a process.

If the organisation is working on the research necessary to determine the long-term positioning, P3, and the senior team has already drawn conclusions on the implications this could have on capability, C3, then the senior management workshop should also take the opportunity to consider the nature of the C2/P2 point on the journey and the route to get to C3/P3. However, it is worth re-emphasising a point made earlier that if the positioning work indicates that the level of change required in the longer term will need to be radical and innovatory, then bringing this into the short-term implementation planning at this stage may only serve to confuse people as they will have difficulty relating the bigger business themes to their current process problems. At this stage due account of any P3 positioning conclusions can be used to avoid confirming process changes that could make more radical change difficult during Step 2.

Some of the ideas generated during Step 1 will be radical and innovatory but they will be out of context if the work on positioning is not complete. Such ideas will need to be brought back into Step 2 for inclusion, or otherwise, at that time.

Implementation

The implementation of changes across a process will be vulnerable to latent functional parochialism and the weakest functional link in the process chain will determine the pace at which the organisation will benefit from the changes. To confirm the change in business perspective, the senior management workshop should also be the vehicle to debate and set up Process Owners and Process Improvement Teams. The latter are a necessity to ensure that all the change proposals are co-ordinated and actioned in the priority determined by the workshop. On their own, their role will be temporary, ending with their return to their respective functions. However, in combination with nominated Process Owners, the organisation will take on a different characteristic, where the needs of functions become subordinate to the needs of the processes. The Process Improvement Teams then become part of the ongoing management framework within which continuous process improvement takes place.

Fitting all the stages of Step 1 together

Assuming that an external customer needs survey has shown that a positioning gap, P1 to P2, has to be closed then the sequential stages of Step 1 are:

Stage 1: Define the current capability:
- Process mapping;
- Data collection;
- Core/support/diversionary time analysis;
- Questionnaires/staff surveys.

Stage 2: Propose options for change:
- Process improvements;
- Changes in service levels.

Stage 3: Review impact of change:
- Internal customers and suppliers.

Stage 4: Decisions to change processes:
- Change process/methods/systems;
- Levels of service;
- Reduce/increase/rebalance resources;
- Implementation plan.

Each stage takes around the same elapsed time. However, the overall elapsed time will be a function of:

- the type and geographic location of any external customers that are involved in a customer needs survey;
- the deployment method used to communicate the need to start the BPM journey;
- the overall numbers of people in the scope;
- the degree to which support team facilitators can be released full-time from the organisation;
- the number of other change initiatives that are running concurrently.

Experience suggests that Step 1 cannot be undertaken sensibly in less than three months, where this short time scale would be driven by a serious business imperative, typically, drastic cost removal. A more considered approach that allows sufficient time to raise and debate proposals and start the process of management behavioural change will need six to twelve months.

The positioning research can start at any time during Step 1. However, the conclusions from this work may point towards a more radical and innovatory series of changes that should be reserved for Step 2. An early introduction of the positioning conclusions during Step 1 is likely to confuse staff who will still be involved in creating the firm foundation of improved current processes.

As the organisation starts implementing the changes arising from Step 1, then the support team of facilitators can begin to introduce additional tools and techniques of continuous improvement.

CONTINUOUS IMPROVEMENT

There are different types of change to business processes. By plotting changes in capability in relation to time, as shown in Figure 3.21, we can define the types of change:

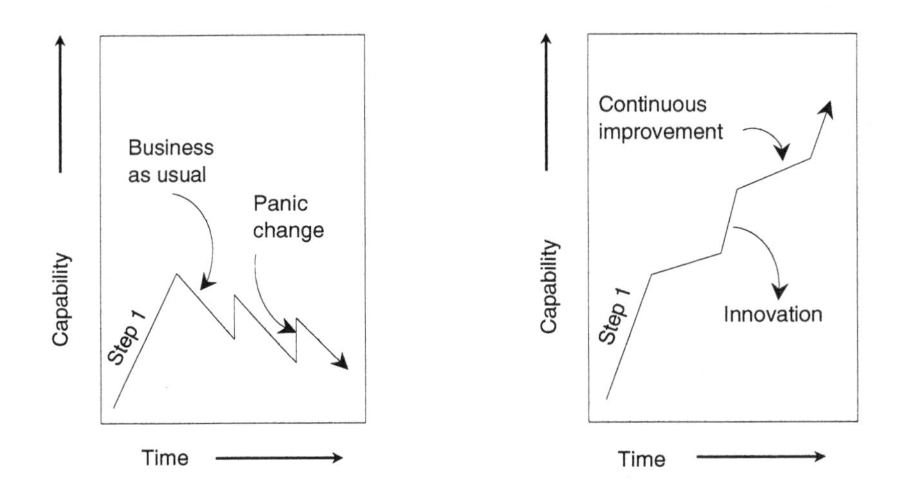

Fig. 3.21 Different types of process change

- *Business-as-usual:* The gradual deterioration as processes fail to meet changing needs;
- *Panic:* The rapid reaction to discovering the business has deteriorated, but without the application of science (e.g. sudden cost cutting);
- *Continuous improvement:* Everyone involved, cross-functional teamwork, process measurement, external and internal customer focus;
- *Innovation:* Radical and innovatory re-engineering of the business to achieve a differentiating customer proposition that leads to business growth.

The purpose of Step 1 was to avoid the panic reaction. Having begun implementing the changes that emerged from Step 1 then there is always the danger that the business will slip back to its old business-as-usual approach with the risk that a panic reaction is used to fix capability issues. Moving to continuous improvement provides the stable base from which to move to Step 2 of BPM.

The transition from Step 1 to a state of continuous improvement is shown in Figure 3.22. As the outstanding number of proposals for change diminish, then the organisation needs to learn the additional tools and techniques of continuous improvement and address the changing role of managers to one of providing support, facilitation and education.

During Step 1, many lessons are learnt and the relationships between departments improves through increased awareness of cause and effect across the

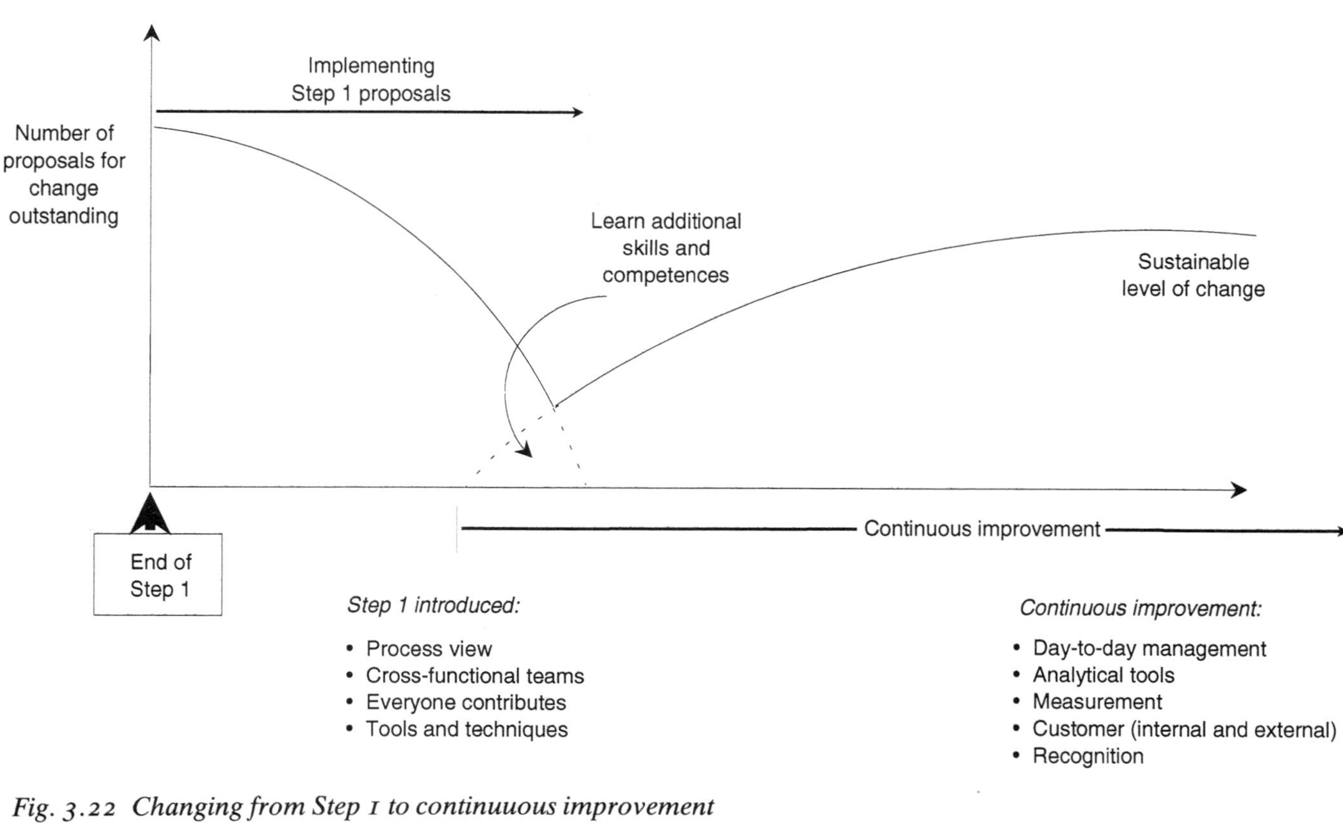

Fig. 3.22 Changing from Step 1 to continuuous improvement

organisation. Everyone and every group will be aware that they have internal customers. It is important to continue to be aware of customers' needs, and as a customer oneself, to ensure that one's own demands are those necessary to meet the overall business objectives and not just to satisfy a parochial need.

It is essential to guard against recurring systematic failure, and the start of new and continuous activities in a group that should be stopped by applying a new conditional change. It is never useful to apportion blame; the opportunity exists to find the root cause and resolve it with other groups in a common process. If the group itself is a root cause of problems, then it will need to concentrate its efforts on prevention of these problems. If the group's uncertainty rises and it finds that it is checking the work of others, then it will know that insufficient prevention is taking place somewhere else. If the group finds that much of its time is spent on correcting work, then the organisation has paid for the work twice and someone has not understood the need for analysis of the process so the process can be improved. Robust processes deliver right first time, not exhortations to people in the process.

Some problems will have been in the business a long time. In general, an intuitive guess at the solution will probably not be the best one. Analysing processes is a science, and Step 1 will be only the start. Some of the tools and techniques of continuous improvement would have been used during Step 1, for selective analysis of process problems. All the techniques will need to be introduced throughout the organisation, by education and training on a continuous basis. This is the route to ensure that the stable state for the company is continuous improvement.

Creating a framework for continuous improvement

Without a framework, managers and staff will drift and the momentum for change will be lost. The overall framework will need to be designed to suit the needs of each organisation, but consideration will need to be given to the following main categories:

- organisation;
- approach;
- education and training;
- communication;
- documentation;
- monitoring and measurement;
- recognition and reward.

It should be expected that the facilitation support team, used during Step 1, will contain team members who will want to change their own career path to one of continuous learning and facilitation in support of continuous improvement for the whole business. The team members' role, at this stage, is to work with the executive team and line management to determine the mix of actions in each category to achieve continuous improvement across the business.

During Step 1, the executive group, support team leader and team members should take every opportunity to visit other organisations that are further along the BPM journey. This helps to avoid some of the pitfalls that lead back to following the old path of business-as-usual, and provides exposure to a broad range of options to achieve continuous improvement. Such visits should not be restricted to one's own sector. The issues are about people interacting with each other and the systems in which they work. This theme is common across all sectors.

Behavioural barriers to continuous improvement

After Step 1, people's behaviour in the light of the need to move to continuous improvement will still be driven by a range of values and beliefs. Just participating in the stages of Step 1 does not, in itself, require a fundamental change in behaviour, particularly in respect of managers. However, Step 1 will expose the range of behaviours, some of which will be carried forward into continuous improvement as a barrier to progress. As always, those individuals with high influence who hold a negative attitude pose the greatest challenge. If the leverage on the business results is high then winning them over will make them key enablers, but sometimes a hard decision to remove them entirely will remain the only course of action.

The relationship between managers and their staff for various management styles is shown in Figure 3.23. The key issue is a change in a manager's

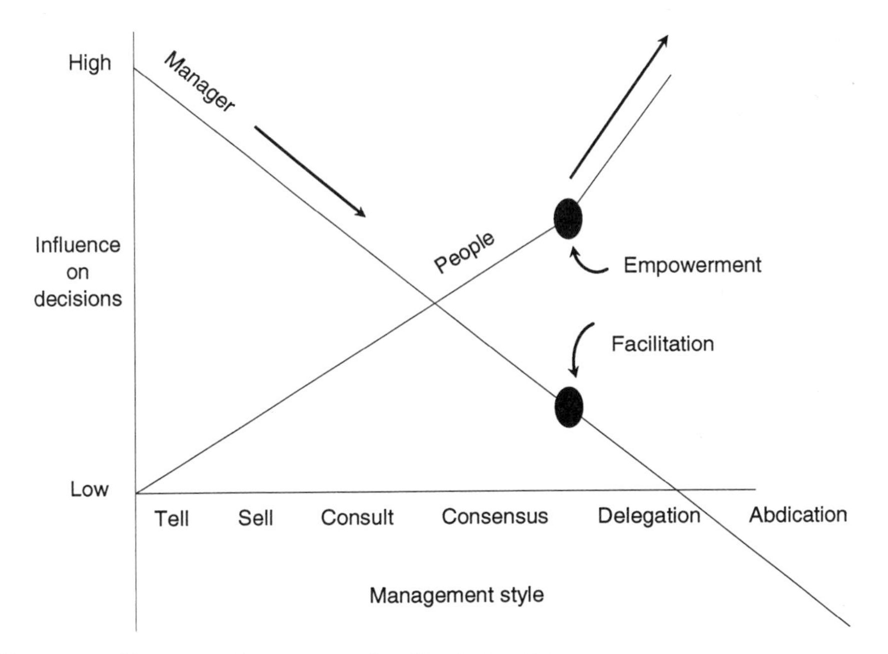

Fig. 3.23 Changing the manager/staff relationships

behavour from instruction and control to help and facilitation. At that point, empowered staff can contribute to process improvement on an ongoing basis. For any individual, a behaviour change will be influenced by a number of factors within the whole organisation. Typical factors include inertia, relevance to the business, lack of resources and lack of critical mass. Of these, critical mass is important in that it is not a function of just numbers of people. The staff will already be predisposed to change towards empowerment as they will have been the victims of poor processes and the victims of ineffective appraisal procedures. However, a small number of the senior team can create the critical mass by their example as the old culture will still have in place the need for middle management and staff to use the senior team as a role model even if in-depth understanding of the need for change is lacking in the middle levels.

Assuming that everyone is doing things in a way they personally believe is the right way, given that they will be motivated to achieve a good appraisal, then we can assume that poor management behaviour is a function of ignorance of the new requirements by which the business will measure them. Creating an environment that exposes what people believe and how people behave, is critical to obtaining change. Doing this in an old culture of fear is particularly difficult and is a reason why the top team should be the first to be exposed to this issue. People can look at each other and at themselves through what is known as the 'Johari Window'. The window is shown in Figure 3.24. The two difficult areas are:

- Unknown to self/known to others: this requires disclosure by one's peers and staff;
- Known to self/unknown to others: this requires disclosure to one's peers and staff.

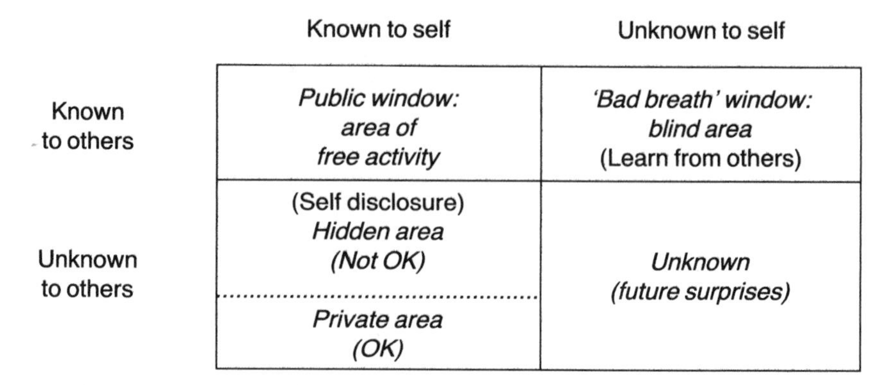

Fig. 3.24 Looking at yourself through the Johari window

Earlier in the chapter, we saw how an employee survey can provide a quantified measure of the perceptions held by everyone in terms of horizontal and vertical relationships. The output from the survey provides an appropriate mechanism to start the process of looking into the 'Johari Window'. The use of the survey will prompt the need to take action on attitudes, beliefs and behaviours. In many ways, using an employee survey can ensure that the issue is tackled and that there is no turning back.

It is important that exposure of poor attitudes and behaviours is not used as an opportunity to vilify individuals – they were only doing their best in a manner they believed was appropriate, for whatever reason.

Using tools and techniques

Knowledge, attitude and skill also apply to the use of tools and techniques to look at processes and solve problems. Few of us are taught how to measure processes, draw conclusions from the data and put in place actions to correct process problems. There is an assumption that managers will know what to do in all circumstances – that is why they are managers. This is re-enforced by the use of the word 'accountable'. People are held accountable for their actions on the basis that there is a presumption of failure. If they do not get the right result then they can be held accountable. However, if they do not have the knowledge then what are they being held accountable for? Accountability is a convenient way of being able to put the blame on someone. People are responsible for achieving success. Who is held accountable for success?

The price of failure for people low in the structure, who are held accountable, can be serious for the individual – no pay increase, reduced promotion prospects, etc. That is how we reward people when we have never given them the knowledge to do their jobs effectively. Those individuals who perpetrate massive fraud or play with the company's financial assets on the exchange markets, or who rig mergers and acquisitions that transgress the law, are truly accountable – they have misused their knowledge and skill.

The key issue is not which technique to apply in which situation (though still important), but the need to acknowledge that processes and process problems can be found and prevented by applying tools and techniques. Traditionally, schools, colleges and business schools have not included the tools and techniques of continuous improvement on their syllabuses. To admit that they are necessary to learn is an attitude barrier in the working environment. The normal approach is to learn the current ways of working from one's peers and managers or to learn from one's mistakes (with whatever consequences). This is not a scientific approach.

In some of the stages of Step 1, a number of tools and techniques will have been used which are summarised in Figure 3.25. For more advanced practitioners, a number of techniques are available with more specific applications. Some of these are shown in Figure 3.26. For a full explanation of their application, specialist publications should be consulted.

Tool/technique	*Use*
External customer survey	To understand the needs of external customers.
Internal customer survey	To understand the perceptions of internal services.
Benchmarking	To compare similar processes to find best practice.
Process flowcharting	To understand how work passes between functions.
Affinity analysis	To measure the strength of functional relationships.
Service performance	To quantify the importance/performance of services.
Activity data	To understand the allocation of time in processes.
Activity categories	To obtain the level of core/support/diversionary activities.
Activity drivers	To relate volumes of activity to causes.
Staff survey	To obtain employee perceptions to work environment.
Brainstorming	To generate ideas for improvements.
Cause and effect diagrams	To prompt ideas during brainstorming.
Scatter diagram	To view the correlation between two variables.
High–Low diagram	To group objects using two variables.
Bar chart	To plot the frequency of an event.
Run chart	To show how a variable changes over time.
Pie chart	To show proportions of a variable as segments.
Force-field analysis	To show the forces acting for/against a variable.
Histogram	To show frequency of a variable in a range.

Fig. 3.25 Tools and techniques of continuous improvement

Tool/technique (advanced)	*Application*
Statistical process control (SPC)	SPC is a means to understand if a process is producing and is likely to produce an output that meets specification within limits.
Failure mode and effects analysis (FMEA)	FMEA is a means to understand the nature of potential failure of components and the effect this will have on the complete system.
Quality function deployment (QFD)	QFD is a structured process to build customer needs into the design of products and manufacturing processes.
Taguchi methods	The design of experiments to create robust processes/ products where final quality is subject to many variables.

Fig. 3.26 Advanced techniques

A SUMMARY OF THE BPM JOURNEY SO FAR

If a company has decided to make the full BPM journey then it will need to plan the stages and timing to complete Step 1. The stage of the journey is shown in Figure 3.27. The degree to which timescales are compressed or relaxed will be a function of economic necessity, desired accuracy and required employee perception. A need to implement a fast downsizing of the business will lead to a limitation on the level of detail that can be explored and any subsequent reduction in employee numbers will leave the remainder less willing to participate in continuous improvement. Good communications and honesty will help avert a poor employee reaction to the BPM initiative. A longer-term approach is always desirable as it allows more time for detail and involvement and a greater degree of re-balancing of the resources through creating spare capacity to work towards growing the level of business.

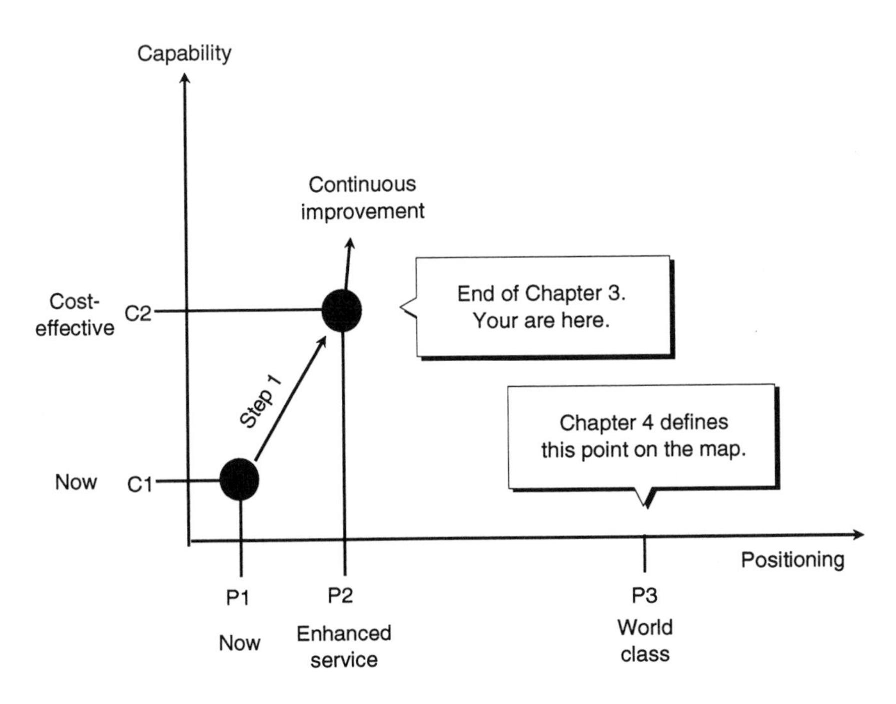

Fig. 3.27 The BPM journey so far

Step 1 presents an ideal opportunity to look closely at everything that is done and question its relevance to the current business objectives and current customer needs, and the impact these have on the changing roles that people will need to have in the future. The outcome of Step 1 is determined by everyone in the process, after all they are the best people to propose what the outcome should be.

After Step 1, the organisation will stabilise on the P2 position. Subsequently, its capability will be changing through continuous improvement, but the level of change will create only marginal improvements to P2.

The next chapter outlines the work required to understand a company's future positioning, P3. Depending on circumstances, the work on positioning to define the point P3 may have started during Step 1. Some conclusions may already be emerging that point to the need to consider more radical and innovatory approaches that lead to business re-design and a greater degree of corporate transformation.

4

CREATING LONG TERM FUTURE POSITIONING

'Cheshire Puss,' Alice began, rather timidly . . .' would you tell me please, which way ought I to go from here?'
'That depends a good deal on where you want to get to,' said the Cat.
'I don't much care where –' said Alice.
'Then it doesn't matter which way you go,' said the Cat.
'– so long as I get somewhere,' Alice added as an explanation.
'Oh, you're sure to do that,' said the Cat, 'if only you walk long enough'.
Lewis Carroll

STANDING STILL OR GOING NOWHERE

In Wonderland, Alice had a strategy that took her to many exciting and unreal places. The Cheshire Cat gave her sound advice appropriate to Alice's questions. For businesses, just going somewhere and expending much energy in the process is a recipe for disaster.

A crisis

In Japanese the word 'crisis' is made up from two characters that stand for threat and opportunity. The recession of the early 1990s posed a significant threat. Alert companies saw that this created a unique set of circumstances which they could exploit. While competitors faced downturns in business volumes, pressures on profitability and strategies that were based on cost-cutting, the winning formula was to get into a position to implement a proposition where success brought:

- a growth in total market share, without having to buy it by price-cutting;
- a growth in the total market, by bringing more customers into the sector.

Companies achieved this by:

- changing the relationship with customers to achieve customer retention and advocacy;
- creating access points to the organisation that eased the buying method and

locked out competitors;
- becoming the lowest unit cost supplier among its competitors, and therefore achieving permanent price competitiveness.

Old Strategies

Many company strategies in the past have all been similar and relied on treating the customer as an asset – an asset to be milked. This essentially works in a rising market. The equations break down in a constantly falling market as revenues decline. Three effects occur:

1. Costs are cut, and often these are activities to do with building the future. Everyday transactions need to continue and companies are also burdened with arrears, bad debts, and the additional processing costs.
2. Staff, particularly in the selling operations, become the victims of the equation. The high-level business ratios are converted into sales targets. In the field, often little can be done to influence sales, but targeting staff presumes that they can. They can influence sales to some extent, but mainly by pressure selling of products and services that are not in the interests of the customers. Such practices catch the interest of the press and become the source of negative public relations for each sector and specific companies within it.
3. Companies, traditionally seen as being trustworthy, suffer pressure on margins and profits which leads them into using questionable practices on a regular basis. The wary customer is now making more reasoned choices.

If this list is the consequence of old fundamental bed-rock strategies then such strategies are now seriously flawed as they will not deliver a future for those that practise them.

Customers not assets

Most companies recognise that customers are important and that keeping them is a profitable route. However, most 'Customer-first' programmes only address simple elements of the service and do not address the fundamental flaws in the strategy. As a result, most customer-first programmes deliver very little. Where companies have recognised that long-term customer relationships, through retention, are valuable, then they take a more serious approach to making changes. For most companies this has meant starting a fundamental re-think of the business. If a company's competitors are all active in this field, then a response is needed just to keep up. The proposition for the future has to overtake the others and sustain that position.

Influences on the customer are changing

The key influences emerging in the 1990s, mainly changed as a result of customers' experiences in the 1980s, are some of the key drivers that companies need to respond to. Many industries are now in a state of flux. Radical change is needed and companies have been slow to recognise what are irreversible changes in the marketplace. As a result there are too many players, and general dissatisfaction with service and customer relationships.

The hard lesson that organisations must learn, is that it is not enough to manage their own businesses. To be successful providers of products and services will have to be more knowledgeable and responsive to customers to an unprecedented degree. These issues have a direct impact on value and pricing and ultimately on profitability overall. Many strategies have been to survive the recession. This has led many to look backwards and inevitably the focus has been on managing the problems inherited from the late 1980s.

Demographic shift inexorably changes the shift in consumer needs based on the spending power that large numbers of emerging age groups will have. On a larger scale, whole populations in one decade will be influenced by a previous decade, creating forces such as income polarisation (the have and have nots) and long-term financial shifts in spending power (the indebted society).

Consumer booms, fuelled by rising house prices, and households taking on fresh debt, lead to a situation where the amount of debt held by the personal sector rose sharply as a proportion of annual income. Economic theory cannot show what level of debt the personal sector will wish to hold but if the debt–income ratio falls back, then, for example, this would raise the spectre of falling house prices over a long period with economic consequences to many consumers and the suppliers to them of products and services.

The competitive threat will come from many sources

Traditional competitors re-focus. Banks offer mortgage products, building societies become banks. Companies founded in a traditional marketplace of agricultural fertiliser products are now generating over seventy-five per cent of profit from consumer pharmaceuticals. Estate agents are moving from their core activity of selling houses to offering a complete financial service. Retail goods chains are offering financial products, petrol stations are turning into supermarkets, and supermarkets are offering highly discounted petrol. New access methods, based on emerging developments in information technology, are offering customers time savings and different ways of accessing services. Direct-line access to insurance and mortgage products, and twenty-four hour banking services are setting new standards of customer service and changing customer perceptions of how to access suppliers.

The implications will be profound as the intensity of competition becomes fierce from massive over capacity. This will lead to substantial numbers of closures, mergers and acquisitions, and weak players squeezed out.

An opportunity

The opportunity exists for the alert organisations to build a strongly differentiated market position for the late 1990s and beyond. To stand above the competition is achievable, but will take time, leadership and a radically new approach to positioning and capability. In a competitive sense, if your competitors are all active in this field then the organisation must respond just to keep up. To get ahead, and stay there, means embarking on more fundamental change. Until recently, many organisations had no sound proposition to make this happen. On the contrary, the previous traditional strategies have had just the opposite effect.

Pooling resources

Developing the future positioning of the business is a science. For many businesses, this activity is a poorly co-ordinated mix of outputs from the marketing, strategic planning and finance departments, the whole often usurped by a range of parochial functional plans from the operations departments. For many organisations, it is not always the case that the appropriate data are missing. Rather, it a failure in the mechanism of bringing all the data together in a way that allows the appropriate conclusions to be drawn.

From customer and competitor research, the key positioning conclusions create the vision, expressed in terms that will delight customers and turn them into advocates of the business, and point the direction to change capability. Through a process of deployment, every employee needs to see the same vision, but expressed in a language they understand and so they are aware of how their own element of the process can contribute to making the vision a reality.

The BPM journey at this point requires that the organisation:

- collects the appropriate data from which to derive the future positioning (P3);
- has the ability to understand the implications of the data to visualise the necessary radical and innovatory changes to capability (C3).

Actions to determine positioning draw on a range of techniques. Positioning starts with the customer and the competitive environment in which the company is operating. If you get it right for the customer, relative to your competitors, then the financial performance will follow. However, the key financial ratios within organisations cannot be ignored. External forces that come from shareholders and/or City watchers will be bringing pressure to bear on the company. Internally, a certain rate of return will be needed to fund growth and investments.

This chapter is concerned with those actions that must be considered in order to collect the data from which the future positioning can be derived. For some organisations, not all the analyses will be appropriate. For others, certain specialist information will be required in addition to that shown here. The key is to ensure that the top team treat deriving the future positioning as a key process in its own right and do not create a number of future positions based solely on each function's perspective. As a minimum, the positioning work should consider the following aspects:

- Macro-economics, industry and competitor analysis – financial scenarios, threats and opportunities.
- Customer and product segmentation and profitability – process costs and revenue.
- Image and attitudinal tracking – societal and customer trends.
- Branding and brand positioning – attraction and retention.
- Distribution and channel modelling – relative access points.

For ease of explanation and to provide a consistent link between all the analyses, this chapter features a fictional finance company that has decided to build up an overall picture of its future positioning. Although focused on the finance sector, the analyses (and conclusions drawn from them) are appropriate to many sectors. Although many of the trends and analyses are drawn from actual data, it is essential that any company undertaking positioning research should validate the data and their relevance to their own business. The conclusions which are drawn out of the data in the example are an amalgam of a number of issues from a range of sectors and are therefore generic in nature.

MACRO-ECONOMIC TRENDS, INDUSTRY AND COMPETITOR ANALYSIS

The macro-economic trends will affect every sector to some degree. Although there will be Black Mondays, Black Wednesdays and other single point events, the main underlying trends have long-term cycles lasting five to ten years. Companies need to be ever alert to such trends, particularly if their products or services are geared to a particular trend direction. Where product development or service delivery changes take some time to be fully in the marketplace then macro-economic effects could be changing out of step with the company's ability to keep up.

Some aspects of the industry and competitor analyses will be generic, such as consumer spending constraints or major societal shifts in service perceptions. Other aspects will be specific to a sector with little impact outside of the sector. To demonstrate the type of analyses required to be undertaken, some of the specific examples used in this chapter are based on a fictional company in the finance sector. The methodology is appropriate for all sectors.

Main economic trends

First, confirm and update the key economic indicators and specific national trends that influence your market. Figure 4.1 shows the key trends in the UK up to 1993 and estimated thereafter. The analyses will need to roll forward year on year.

The range of forecasts will be a function of many variables and can change rapidly. The key is to apply best knowledge to each trend, paying particular attention to those factors that will have greatest leverage on your own organisation. For our example company, the key issues would be summarised as:

1. The scale of outstanding consumer debt that will prevent consumer demand leading the recovery.
2. Unemployment widening the gap between the haves and have nots – arrears and bad debts still a burden.
3. Inflation to remain flat due to depressed employment opportunities.

For some sectors, a European or world dimension would need to be used to track economic trends. For example, a manufacturer of agricultural vehicles would need to track the changes in policy on farm subsidies in each country.

Specific market scenarios

For the particular competitive environment the company is in, a number of market scenarios can be created. Our finance sector example company is involved in lending (personal and mortgages), savings, investments, insurance and current account cash transactions. A sector-specific analysis for mortgages in the UK is shown in Figure 4.2. For our example company, the key issues would be summarised as:

1. Negative equity problems lead to ability-to-pay criteria taking precedence over value of asset in terms of lending criteria.
2. House prices rising at around inflation rate without fast swings in value.
3. Rising then stagnant number of mortgage applications.

Similarly, a trend analysis for the savings market would be completed. For our example company, the key issues would be summarised as:

1. Conventional savings under attack from other equity products.
2. Influence of consumer debt repayment reduces.
3. Rate of increase in balances halved. Further reductions could have serious consequences.

Competitor actions

First start by listing all the traditional players that are in your sector, and have the potential to become greater competitors in the future. For our example, the UK players in the sector are shown in Figure 4.3.

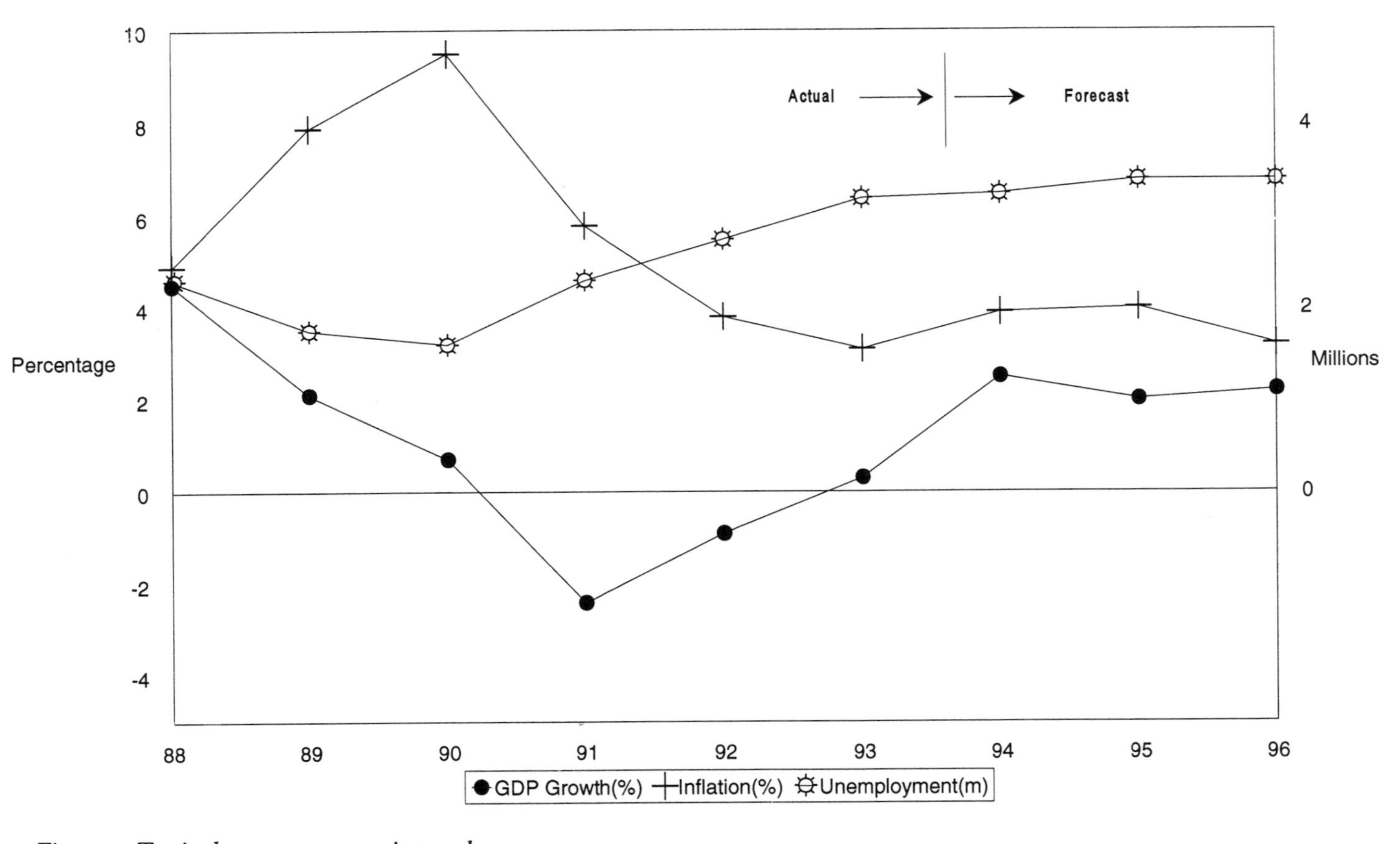

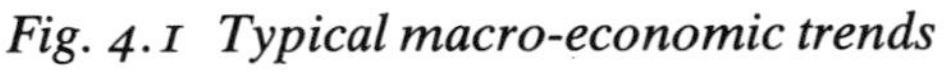
Fig. 4.1 Typical macro-economic trends

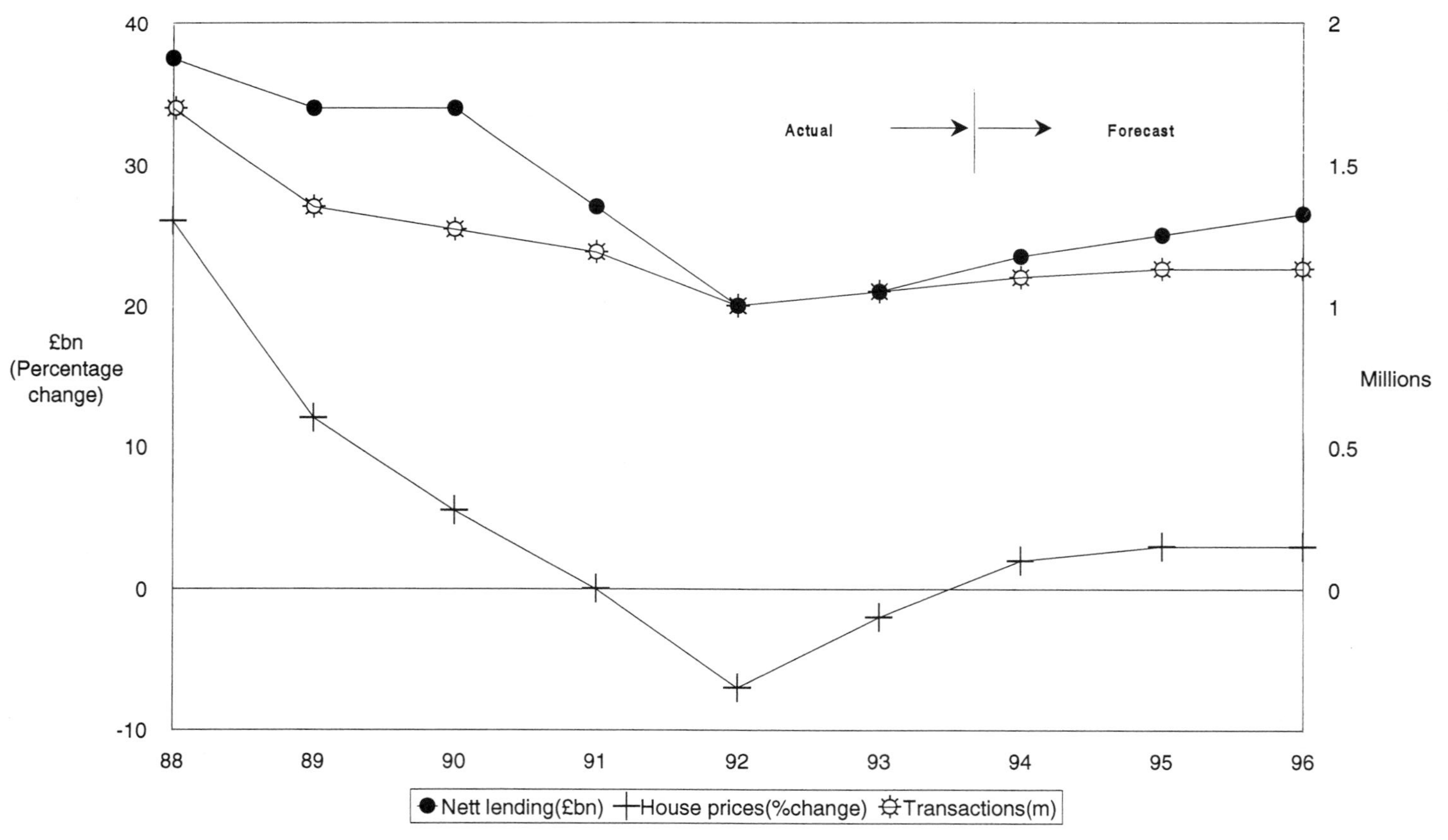

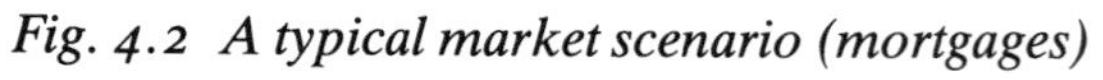

Fig. 4.2 A typical market scenario (mortgages)

Competitor type	*Product/service type*
Goverment	National Savings, pensions, gilts
Banks	Retail, commercial, merchant, clearing
Commodity exchange	Foreign, capital, central lending, retail credit
Building societies	Mortgages, savings, loans
Insurance	Agents, brokers, reinsurers, loss adjusters, Lloyds
Unit trusts	Funds, unit trusts

Fig. 4.3 A typical competitor matrix

This can be subsequently broken down into the constituent elements of products and services, and your own company and the competitors mapped against them.

Taking the overall sector, a number of broad comparisons and trends can be established and the market scenario in relation to competitors derived. For our example company, the key issues would be summarised as:

1. Poor financial results will make more companies vulnerable to mergers and acquisitions.
2. Effects of arrears and bad debts eroded by regressive recovery actions.
3. Cost–income ratios improved by aggressive cost-cutting.

The competitor matrix and overall trends, when combined, provide a basis from which to forecast some of the competitor moves that are likely to be made and some of the potential responses your own business could make in relation to the competitors.

Changes in factors that determine industry profitability

The key factors that determine profitability for any company are the rivalry between competitors, the influence the service provider has over customers, the degree to which customers can impact the business, the threat that new entrants to the market pose, and the range and variety of competitor products and services in relation to traditional offerings. This exercise should be undertaken as a trend analysis in order to establish the rate of change in these factors and to assist in predicting likely scenarios in the future.

For our example company in the finance sector, a typical analysis would expose the trend shown in Figure 4.4. The trends for the 2000s indicate that they will apply generically, not just to the finance sector.

Factors	*1970s*	*1980s*	*1990s*	*2000s(?)*
Rivalry	Cartels, co-operation, legislation to keep apart	Cartels, break up, margin pressure, cross-territory	New entrants, margin erosion growth slows	Mergers and acquisitions, only the best survive
Service provider influence	Rationing, regulation, product/supplier separation	No rationing, deregulation, greater choice	Fight for market share, customers recognised	Only through delighting customers
Customer influence over supplier	No choice, loyalty, unsophisticated	Erosion of loyalty, awareness	More choice, high service expectations	Loyalty through partnership, not price alone
Range and variety of competitor products/services	Few alternatives, single sourcing	Shop around on price, expansion from deregulation	High sophistication to match need, channel choice	Differentiation through relationship, not channel

Fig. 4.4 A typical trend analysis

New entrants

A constant threat for many companies in all sectors are new entrants into a market. The particular threat comes when the traditional product or service provider is limited in the range it offers but the new entrant competes directly by adding an additional service to its non-competing range. This trend is noticeable in the retail consumer goods sector where financial services are tagged on (store cards, personal loans, investment products). The quality of the retail chain's brand image transfers to the new service, whereas the traditional suppliers may be suffering from negative public relations directed at the whole sector. Banks selling fresh farm produce and clothes off the peg is not an easy counter strategy to implement.

Suppliers linked in a process chain build separate core services. However, the organisations that are earlier in the supply chain begin to provide the downstream services and cut the chain of supply short. For example, Estate agents who obtain the mortgage and insurance for their customers then keep the customer relationship rather than pass the customer down the supplier chain. The emergence of natural gas as a raw material for power generation allowed new entrants to emerge while traditional generators based on coal still struggle to get a return on their original and higher investment.

Deregulation allows new entrants to capitalise on emerging consumer trends that would not be satisfied by the traditional operators. Bus deregulation created a new type of small vehicle with a service that stopped anywhere along a street on demand. Traditional vehicles were not suited to this approach.

A new entrant can start with an insignificant attack on a segment ignored by the traditional supplier and then move through the segments to make a full attack once the brand image has been established as competitively superior. The four-wheel drive off-road vehicles that emerged from the East were designed to be attractive to image buyers, rather than serious off-road users. Having created a new market for such vehicles, later derivatives then attacked the traditional user market, leaving the original supplier without an immediate response to attack back across the whole product range. Whole sectors have been decimated or eliminated by this approach.

For our example company in the finance sector, the key threats can be summarised as:

1. The entry of the retail sector as a supplier, coupled with the convenience of access.
2. The management of the total financial supply chain, and the ownership of the whole customer relationship.

New technologies

New technologies, particularly the impact of IT on delivery mechanisms, can change customer expectations and leave traditional suppliers running hard to catch up. In the finance sector, the emergence of twenty-four hour telephone banking or insurance services not only changes the nature of that sector but can influence how the consumer will want to deal with other service providers in the future. Cable services, either through dedicated lines or the current telephone network, coupled with the high home ownership of computers, will bring a new dimension to buyer choice and product delivery.

For those companies taking the route of differentiation through technology there is always the risk of not capturing the world standard. Video technology started with two standards, now there is one world-wide standard. A multiplicity of mobile phone services battled for some time until only a few service providers remained. The digital versus analogue battle for a new magnetic tape standard, the battle for satellite TV supremacy, the confusion over high-definition, wide-screen TV technologies, are all examples of massive investment with high gain or high loss balanced precariously on consumer choice.

For a while, technology can give competitive differentiation only to have the combination of new entrants and emerging technology improvements force the original suppliers to invest twice. The early 'hole-in-the-wall' cash dispensers changed the nature of cash withdrawal for banks. Later, building societies introduced similar machines with enhanced functionality as well as being able to operate in the traditional banking territory. To catch up, banks need to re-invest to overtake the building societies.

For our example company in the finance sector, the key threats can be summarised as:

1. Retaining the initiative by further investment in technology as a delivery mechanism while still retaining the high asset base of a branch structure.
2. Using technology to assist the process of building a different customer relationship, moving away from a technology base that just supports separate products.

CUSTOMER AND PRODUCT SEGMENTATION AND PROFITABILITY

A company will have the customers it has got and may or may not be able to influence the type and mix for the future. The issue is thus one of knowing which segments the company has and relating these to the trends which influence the customers external to the company. Typically, it should ask:

- if the demographic changes are indicating that the segments for whom the company is attractive are in decline or increasing in size?
- if the segments that have traditionally used the company are now suffering or enjoying a change in financial circumstances as a result of the macro-economic effects?
- if all the customers add value that is greater than the costs of servicing or supplying them?

To answer these questions requires a knowledge of the external trends and, particularly, the relationship between internal processes, activities and customers. The more difficult analysis is that concerned with understanding the costs and activities that are aligned to customers. Computer systems and databases, and particularly the management accounts data, are generally not aligned to enable knowledge about customer costs to be extracted simply.

Spending constraints (all consumers)

A number of constraints on spending have arisen between the 1980s and the 1990s:

1. The fear of unemployment has increased with around twenty-five per cent of white-collar and professional staff to forty per cent of unskilled employed people reporting that they are concerned about job security.
2. Outstanding debt as a percentage of personal disposable income has risen from forty per cent in 1980 to one hundred per cent in 1990, where it is expected to remain at this high level for a number of years. The problem of negative equity has compounded this effect, so reducing spending.
3. In the ten-year period to 1990, the bottom twenty per cent of household disposable incomes fell by eight per cent while the top twenty per cent increased their disposable incomes by thirty-six per cent. The growth in separation has never been so great in any previous period.

4. This income polarisation and other recessionary factors have effectively removed thirteen million out of the twenty-two million UK households that would normally fund an increase in consumer spending.
5. Consumer spending in the 1990s is only expected to rise at a third of the rate experienced during the late 1980s.

These trends will affect many sectors, as well as our example company in the finance sector.

Demographic changes (all consumers)

Depending on the products or services a company provides, changes in demographics can be a boon or signal doom. Figure 4.5 shows the major demographic changes for the UK to the year 2000. The only group showing a growth in size is the fifty-five to fifty-nine age group. The main decline will be in the fifteen to twenty-four age group. The largest proportion will be those people aged sixty and above.

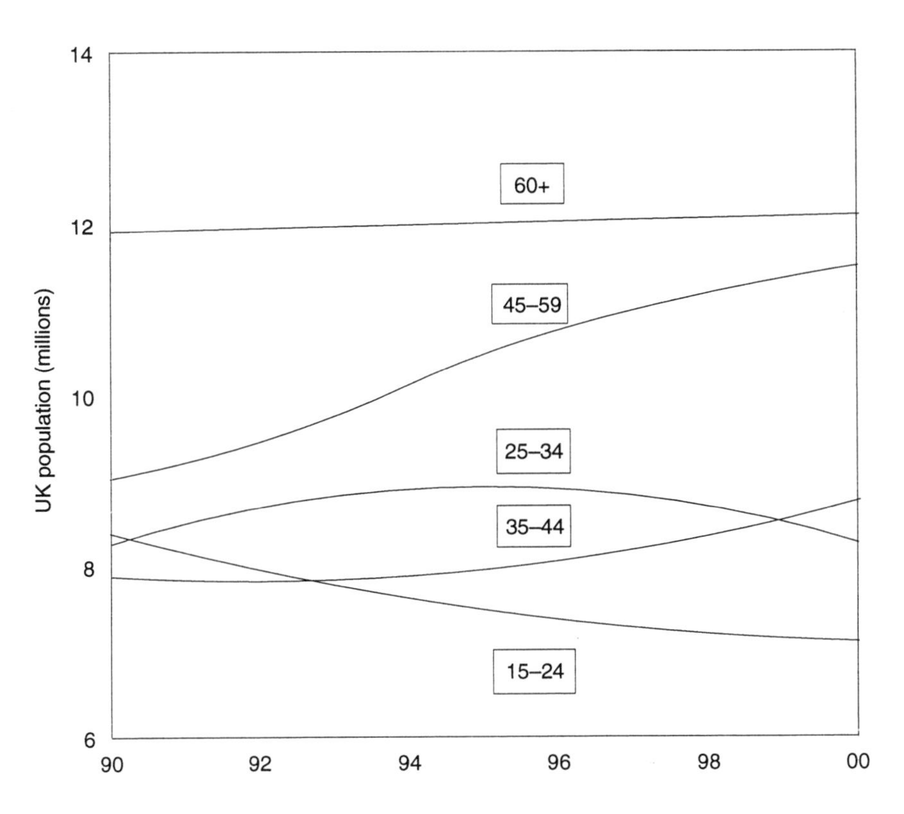

Fig. 4.5 Demographic changes

The implications for our finance company example are that:

2. There will be low levels of new mortgage demand from first-time-buyers.
2. There will be an excess supply of property from the elderly last-time-sellers.
3. Retirement and investment planning will be the largest growing opportunity in financial services.
4. Inheritance is forecast to be a large sector by the year 2000.

Customer segments, actual and desired

Analysis of the company's customer database, backed up with additional data obtained through customer surveys, will bring together a customer's financial attributes and some knowledge of the contribution the customers are making. The results often show a startling result. In Figure 4.6, the analysis for the example company indicated that it was only attractive to one age segment of the population and that a high proportion were not valuable to the business. If this is coupled with the earlier knowledge on demographic changes, then the future shows a declining fortune for the business. In the desired state, the business needs to be attractive to more demographic segments, and within these, the more valuable members.

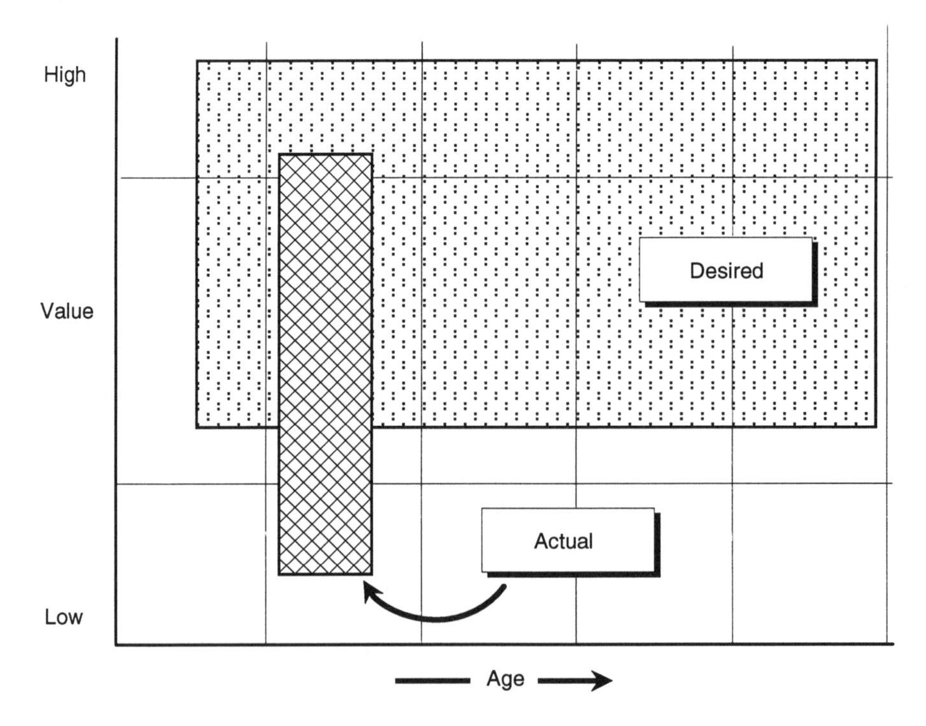

Fig. 4.6 Customer database profile

Where the data are available, then your segmentation analysis can be compared with the competitors. In Figure 4.7 it can be seen that our example company is poorly placed against the competitors. The desired position is to achieve a broader spread of attraction. Although the low-income segment may be unprofitable in the short term, any uplift in the economy would make them profitable. This would also focus on the need to understand the relationship between the process costs to serve this segment to ensure that they could be made profitable through a reduction in process costs.

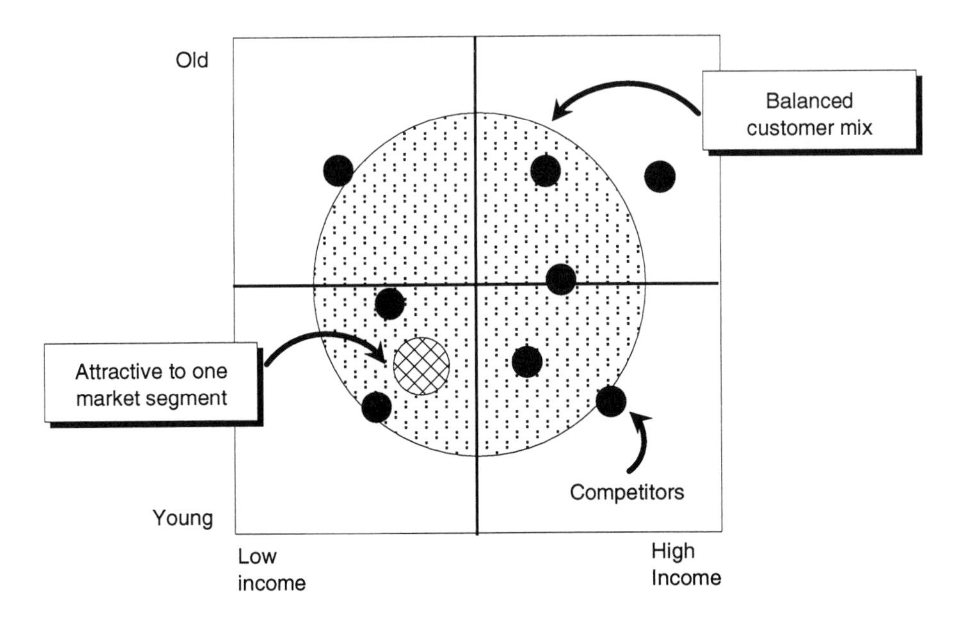

Fig. 4.7 Segmentation competitor comparison

The value of retaining customers

The cost of customers includes the whole process of acquisition, servicing, dealing with customers' complaints, dealing with customers leaving the relationship, and finding replacement customers. If customers are constantly delighted, then a whole series of processes are eliminated. In particular, the cost of finding replacement customers is removed. The value of customer loyalty is high for many businesses and can readily be calculated. Typically for credit card operations a five per cent reduction in defection rate provides seventy per cent increase in the net present value of customers. For an insurance broker it is around fifty per cent and for a software supplier it is around thirty per cent.

The factors that determine customer loyalty are varied and sector specific.

However, the key danger is for a sector not to realise that a new measure has risen in importance in the customers' perception. Over a short period of time, the issue of 'service' rather than price or convenience has become a key differentiator for some sectors. If a company ignores this change then it will be focusing its process improvements in entirely the wrong direction.

Recognising the right segments for your business

Often, segmentation analyses have focused on a number of traditional groups. Consumers have been categorised into A, B, C1, C2, D, and so on. These groups represent particular attributes to do with income, spending patterns, living styles and so forth. Although useful, other types of segmentation are proving to be more significant and are based around an understanding of 'Life Stages'. This type of segmentation works for both consumers and companies.

Company life stages

Companies go through a number of life stages. In an earlier section, we have already mentioned the four major classifications of Entrepreneurial, Marketing, Bureaucratic and Quality. However, another series of life stages can be established to cover the extremes from birth to death of a business. For example, some of the stages could include:

- Formation
- Acquisition
- Consolidation
- New chief executive
- Rapid growth
- Product launch
- Organisational change
- New philosophy
- Merger
- Competitor threat
- Decline
- Liquidation

At each stage, suppliers would have to ensure that they understood the life stage of the customer company, as this could have a bearing on the nature of the relationship with that customer. A new manufacturing and assembly philosophy could change the relationship with a supplier if the change was a move to implement just-in-time. A change to increase the rate of product development could impact on the nature of the IT support the company needed as it moved from transactional-based computing to product simulation.

Consumer life stages

Similarly, applying life-stage analysis to consumers can have a major bearing on how companies respond to market forces:

- how the company develops its brand image;
- the nature of the products and services themselves;
- the relationships staff need to build with customers.

The main life stages are shown in Figure 4.8. The types of curve and particular event are not static. Demographic changes, society norms and macro-economic trends impact on the importance of various life stages. For example, as the incidence of divorce rises and legislation is introduced to change the rights of women and single parent families, then the importance of the divorce rate increases to particular suppliers of services in various sectors. An ageing population of home owners will increase the importance of retirement and death, as this will impact on the higher levels of inheritance of assets.

Our example company in the finance sector found that its mortgage lending division was attractive to people at the life stage of 'First time home ownership', but that at the stage of 'Family formation and move house' over sixty per cent of its customers chose to obtain a mortgage from a competitor. Losing customers at an early life stage also meant that there were few customers to whom the company could appeal at later life stages.

Analysis of a customer database by customer life stage overcomes the vagaries inherent in the A, B, C1, C2, D classifications and the shifts in demographic age groups. However, for a life-stage analysis to be meaningful, the database will need to contain the customer attributes that describe the life stage. Greater subtlety will be needed if the company's service or products are to assist at a life stage transition. The customer attributes or customer relationship then has to have a degree of prediction.

The weaknesses of conventional costing approaches

The demands on businesses to create a new positioning, supported by a competitive capability have increased the overall complexity within organisations coupled with major shifts in the organisational culture and management approaches. Amid all this change, the management accounting practices have generally remained locked in the approaches of the 1950s and 1960s, approaches that do not now afford management the basis on which to make key decisions on products, services, markets and investments.

The inappropriateness and inaccuracy of the traditional methods has now left management unable to answer properly some key questions:

- Why do we so often focus on the short term, rather than the long term?
- Why did we introduce products and services that eventually were seen to have always been unprofitable?
- Why is it so difficult to separate the profitable from the unprofitable customers?
- Why has every cost-cutting programme left us worse off after the event?
- Why did we miss the investment opportunity to attract the profitable segment of new customers which our competitors seem to have attracted?

Where the only regular information that managers receive is contained in the monthly management accounts, then they will fail to receive information concerning the true nature of the internal and external environment in which they are required to make good decisions. Although it is recognised that bad decisions can still be made on good information, the omission of good

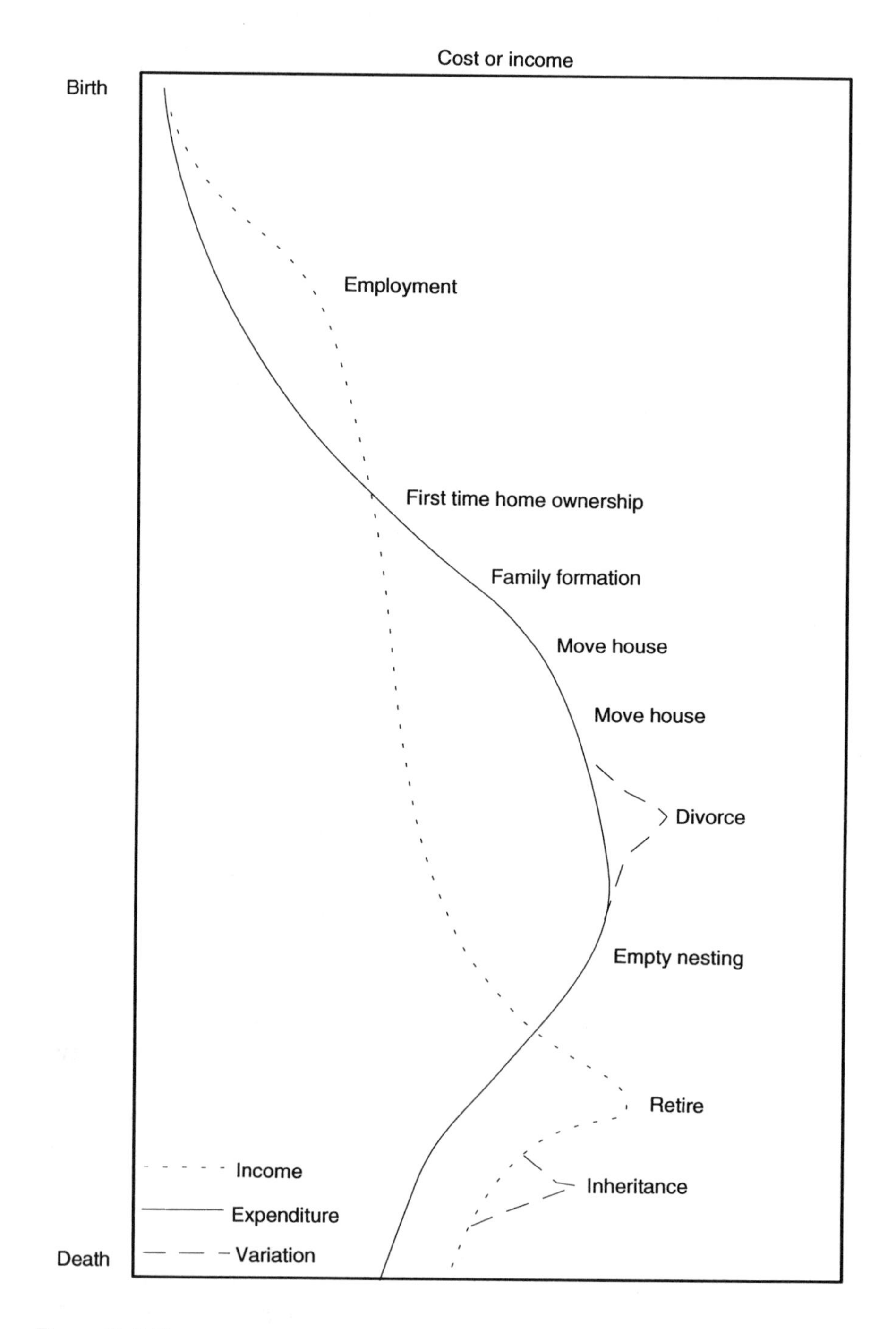

Fig. 4.8 Life stage events

information leaves managers constrained to fall back on their basic intuition and gut feelings.

Conventional management accounts, based on a subset of the financial accounts, have a number of weaknesses that make them inappropriate to the needs of managers involved in BPM. In chapter 3, we also saw how an analysis of activities can provide the costs of processes as well as the alignment of costs to products and customers. In BPM, segmentation has a fundamental link to the costing procedures within an organisation.

The weakness of conventional product costing

A simple 'product cost' consists of:

(Materials) + (Direct Labour) + (Apportioned factory overheads).

The selling price less the 'product cost' is the contribution to general overheads, such as computing, personnel, marketing and so on. Some of these overheads may be in 'cost-centres' and may have been apportioned into the factory overhead area 'cost-centres' or into other general overhead 'cost-centres'. In manufacturing these costs are treated as being 'recovered', based on a volume variable such as direct labour hours. For companies with simple products, low overheads and high labour content, the cost recovery approach works. Any degree of complexity and a shift away from direct labour invalidates the approach.

A company had total manufacturing overheads of 1,500 units and direct labour costs of 500 units. Its overhead recovery rate was therefore calculated as (1,500 ÷ 500) = 300 per cent. This rate was applied to the two products it now made. A conventional cost calculation is shown in Figure 4.9.

	Product A	*Product B*	*Total*
Labour	100	400	500
Materials	400	200	600
Manufacturing overhead	(100 × 300%) = 300	(400 × 300%) = 1,200	1,500
Total	800	1,800	2,600

Fig. 4.9 Conventional product cost calculation

By studying the manufacturing overhead at the activity level, a more accurate indication of the absorption of the 1,500 units by each product appeared, as shown in Figure 4.10. Product A had significant differences from the requirements to make product B. This had gone unnoticed in terms of the

	Product A	*Product B*	*Total*
Labour	100	400	500
Materials	400	200	600
Maintenance	160	40	
Handling	200	100	
Inspection	360	80	1,500
Developing	100	160	
Sourcing	240	60	
Total	1,560	1,040	2,600

Fig. 4.10 Accuracy through activity level costing

conventional cost recovery apportionment reflected in the underestimation of cost 'A' by ninety-five per cent and over-estimation of 'B' by forty-two per cent.

In an example from the automotive industry, a company produced brake assemblies for road vehicles as well as for railway carriages. On closer analysis it was found that over seventy-five per cent of design and production overheads were devoted to servicing the needs of the railway customer although the direct activity and materials were only ten per cent of the total. Pricing had been calculated on direct costs with a fixed overhead recovery rate. Increasing business from the railway company nearly brought the supplier to its knees.

Complexity of design and an increase in manufacturing automation have significantly changed the balance of direct to indirect activity, all of which have created room for major inaccuracies to appear in conventional overhead recovery calculations.

The weakness of conventional customer costing

Overhead costs, such as marketing, distribution, selling, administration and corporate overheads are often treated as fixed and as being recovered as a percentage of sales revenue. In other words, each unit of sales contributes equally to such costs. However, in a case where identical sales volumes to two customers produced widely different levels of sales revenue, the lower sales revenue would be taken as contributing less to the overheads. If the same customer had managed to negotiate a very competitive price and also had higher demands in terms of ongoing service, then clearly this customer should attract a higher proportion of the supplier's costs. Conventional customer costing has the equation the wrong way round.

The weakness of conventional budgeting

Conventional budgeting ignores the impact of each department being in a multi-functional process. There are two consequences of this:

1. The cost drivers in one department are not taken into account by another department. For example, the level of purchasing of parts may be driven by the rate of change in launching new products or in the rate at which design action is needed following an unexpected rise in warranty claims by customers.
2. Many of the items budgeted in one department are treated as allocated cross charges from another department and, as such, are outside of the control of the manager.

Where a manager can only rely on past expenditure figures as a guide, then the fall-back option is to extrapolate from the past into the future. This simple procedure overcomes the need to plan the future or understand the true nature of the activities undertaken within the department and the resulting impact on the business of the activities. The budget thus becomes a series of planned inputs (resources) without any statements concerning the nature of the outputs (the results of activities). Managers are then often judged on the level of resources they have used rather than the results of the activities they have controlled or directed.

In summary, budgets have little to do with the planned activities of a department and the consequences of the activities on other departments, the business or the customers. As such budgets are useless as a tool to improve overall business effectiveness.

The weakness of conventional management accounts

So far we have considered BPM as a journey to the vision. Can management accounts help? The budgets can be considered as a timetable with eleven arbitrary stops along the way of a route that has not been defined. The monthly accounts tell you how far you are from each arbitrary stop but do not say whether the stop was the correct one, why you have missed the stop or how to get back to it or why you went wrong in the first place. Finally, you have to decide where to go next by being constrained to make the journey to the next arbitrary stop by only looking in your rear-view mirror.

A plan to get somewhere, a given level of costs due to a set of planned activities and a given level of revenue resulting from the outcomes from the activities, will have points that can be measured. Being on plan is thus meaningful, as are the consequences of deviating from the plan. Also, measures of the processes will indicate how well the processes are delivering the expected outcomes and will point to the root causes of problems that are creating process failures. Activities have characteristics – some add value, others are diversionary – and activity analysis leads to process improvements. Measurement and proper allocation of costs through a knowledge of cost drivers leads to a meaningful understanding of product and customer costs.

All such measures have time cycles that are relevant to the variable being measured. Some measures are one-off, others as required, while others will be cyclical over hours, days, years. Any system that only accepts measurements on the basis of the passage of the earth round the sun divided by an arbitrary number of twelve would appear to be an oversimplification of the issue and nothing to do with running or improving a business.

Using an activity based cost (ABC) management approach

The ABC approach uses the principle that activities use resources whereas a conventional costing system describes the resources supplied. Conventionally, the difference between budgets and actuals is treated as a variance and much management activity is spent determining the nature of variances. This emphasis leads to judgements of managers on their ability to meet budgets, and a complacency when they do. The principle of ABC is based on knowledge of activities, the reason for their existence and the factors that drive them. By assigning 'pools' of activities (those activities influenced by the same cost drivers) to 'objects' (such as products, distribution channels, customer groups), the ABC model then becomes a dynamic reflection of reality.

Determining cost drivers

To determine cost drivers requires an ongoing understanding of the dynamics and complexity within the business. These variables change over a period of time and need to be taken into account in any ABC model. For example, the introduction of new ranges of components brings with it a series of specific activities associated with procurement, setting up inventory control mechanisms and so on. Conventionally, the implications of introducing a new part, rather than use an existing one, are not known from the accounts. In ABC, the rate of introduction of new parts would be a 'cost driver' and the 'pools' of activities around the company that are influenced by this rate would be linked in the model. With this knowledge, the true costs to the company of new parts could be used to influence design decisions. Without this knowledge, product or service proliferation can become the norm with the overheads that this action attracts being lost in the global reporting systems.

In chapter 3 we saw how, during Step 1 of BPM, the work groups in the organisation listed their main activities. Through discussion, a list of cost drivers can be allocated to the activities. The support facilitation team would initially attempt to create a list of cost drivers to reduce the potential variety in the list. A typical set of cost drivers for a manufacturing business could include:

- Volumes of: materials used, labour hours consumed, parts produced;
- Number of: new parts, new suppliers, new prototypes;
- Number of: customers on file, orders raised, invoices sent, batches scheduled;
- Number of: design modifications, customer warranty claims.

The database of activities can then collect all the activities together into 'pools' having common cost drivers. Changes in the level of the cost driver then relate to the level of activity.

By further breaking down the cost drivers by 'objects' (customer groups or products), a proper allocation of product and customer costs is then derived. Comparisons with conventional overhead recovery procedures will then highlight the anomalies that arise with conventional costing approaches.

Building the ABC model

To build up activity based product costs, the total for any one product (or similar group) would be the sum of:

$$\text{Activity based product costs} = \begin{cases} \text{Direct material \& labour} + \\ \text{volume dependent overheads} + \\ \text{variable cost driver dependent overheads} \end{cases}$$

To build up activity based customer costs, the total for a customer (or group) would be the sum of:

$$\text{Activity based customer costs} = \begin{cases} \text{Activity based product costs} + \\ \text{volume dependent customer costs (e.g.} \\ \text{packaging materials)} + \\ \text{variable cost driver dependent overheads} \end{cases}$$

A final category of costs are known as infrastructure sustaining costs. Such costs have only a weak link to particular product or customer groups. Some types of space cost and corporate overhead fall into this category. For the purposes of making, say, product rationalisation decisions, or determining pricing decisions to make customers profitable or to eliminate them, such infrastructure costs can be ignored. However, if a true calculation of product or customer profitability needs to be established, then using the activity data and a meaningful allocation of non-people costs would need to be completed. There will always be a number of meaningful cost drivers that are better than guesswork or using the conventional overhead recovery formula.

The difference between the conventional and ABC approaches is shown in Figure 4.11.

Using ABC for positioning decisions

The understanding of product and customer costs that comes from using ABC provides its real value when the revenue resulting from the total activity within the business is related to the costs of achieving that revenue. ABC then provides the basis to understand product and customer profitability and allows management to make decisions both on positioning and capability. In other words, the company can determine which products, markets and

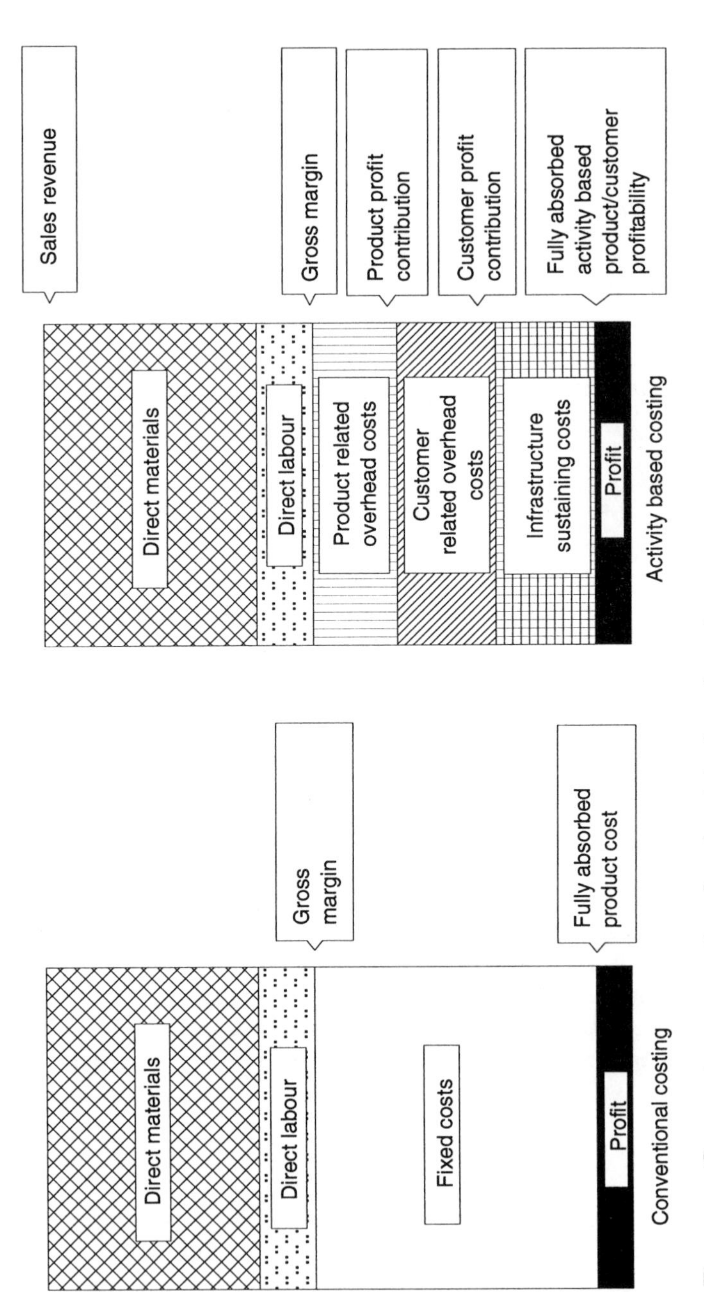

Fig. 4.11 Comparing conventional and activity based costing

customer segments it wishes to retain, it can make relative pricing decisions based on competitor initiatives, and it will be able to focus on those aspects of capability which leverage profitability through process improvement and a reduction in unit costs.

A useful analysis when considering positioning in relation to customer segments is to plot a graph of cumulative customer contribution. A distributor of a mature product range knew that it was profitable at the gross margin level and believed that its complex volume-related discount-pricing structure across a wide range of customers was delivering profit from every product at every outlet. As the warehousing and distribution costs were large the company decided to investigate alternatives to its supply network. Using an ABC model to understand each customer's contribution, the graph revealed a surprising curve, as shown in Figure 4.12. The total contribution from the least profitable sixty per cent of customers was zero, and the least profitable twenty-eight per cent of customers had a servicing cost far greater than the gross margin they generated.

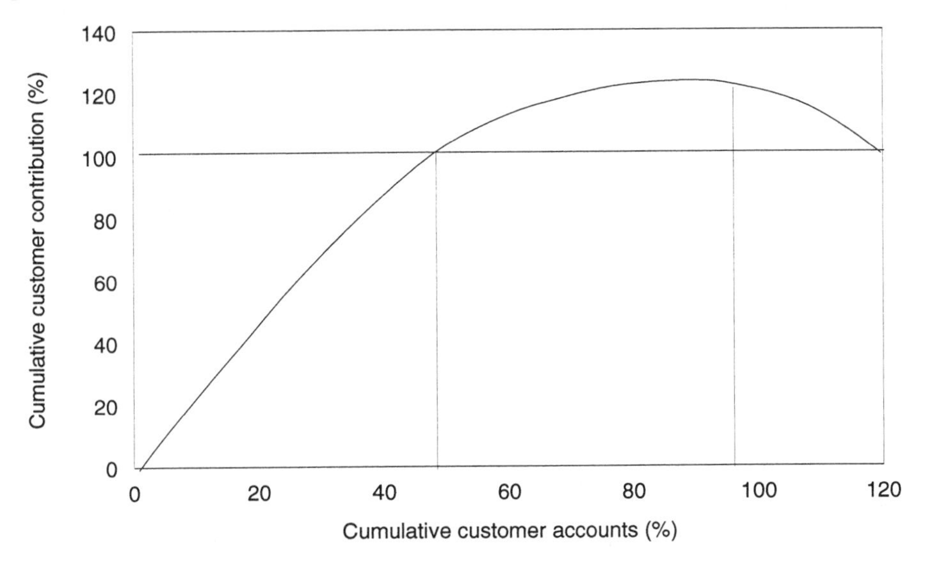

Fig. 4.12 Customer profitability analysis

In the least profitable category, one customer accounted for twenty per cent of turnover as well as a large negative contribution. Tackling a household name retail chain to increase prices is an action for the foolhardy. However, during the analysis of Step 1 in BPM, the incompatibility between the customer and supplier processes accounted for a significant amount of unnecessary work. By working together, both supplier and customer changed their processes to the benefit of both parties. The supplier retrieved an unprofitable relationship and the customer's overhead costs reduced.

Emboldened by this action, further analysis of the distribution network brought to light a significant opportunity to rationalise and reduce costs by persuading a number of unprofitable customers to switch to a competitor. Some remaining customers in the same region could then be serviced more profitably from the upgraded network created from the release of capital after the sale of a number of distribution sites.

Using ABC during the positioning work provides an opportunity to undertake customer-engineering as an action to guide changes in capability. Avoiding activity may then be an initial action rather than spending effort on improving what could be an unnecessary process. As a minimum, the ABC analysis of the current business, prior to making changes resulting from Step 1 of BPM, can answer a number of questions:

- Can the salesforce be directed towards more profitable segments in relation to the costs of servicing the business that is generated?
- Will a switch away from just volume selling to focused selling improve short-term profitability?
- Can some customers or product lines be dropped and achieve an increase in company profitability?
- Can we better apply discounts to achieve volume and know that we remain profitable?
- Can we switch some customers to wholesalers or other third parties rather than service every customer ourselves?

When adept at applying ABC models in one's own company, knowledge of how competitors operate will allow an ABC model of the competitor to be built up. With this knowledge, points of vulnerability will be found that can be exploited through changes in your own pricing which you will know cannot be repeated by the competitor profitably.

ABC makes the inextricable link visible

ABC is a way of linking a company's trading relationships (positioning) to its internal cost structure (capability). The basic premise is that activities consume resources and convert them into products and services to customers. Costs are therefore the consequence of resource decisions, and income the consequence of linked activities – the business processes. The requirement is therefore to improve resourcing decisions as the means of managing costs, and to improve processes as the means of improving business effectiveness and thus customer service and revenue.

The basis of an ABC framework is shown in Figure 4.13, whereby the management framework uses ABC data as a source to support:

- profitability management, such as costing and profitability analysis, customer and product mix decisions, and support for marketing decisions;

- resource and performance management, such as resource to volume and service level changes, activity budgeting and cost driver analysis.

The improvement framework uses ABC data as a source to support both incremental improvement, Step 1, and radical and innovatory change, Step 2.

At the start of BPM, using ABC is another way of making the inextricable link visible within the business and is a contributory element to plotting the BPM journey. The management and improvement frameworks are mapped on to positioning and capability as shown in Figure 4.14. As the BPM journey progresses through Step 1, process changes will be made and unit costs reduced. ABC provides a powerful mechanism to help focus on capability changes while understanding the implications to positioning that such changes will make. Having made the changes, the ABC model can be updated to reflect the new business.

The changes, and reports generated from the data, will be part of the implementation plan. The frequency that the data and reports will be needed on an as-required basis to support management decisions. This frequency is unlikely to be fixed or on a short time cycle. Given the real knowledge and value from using ABC, the adept practitioner will soon be advocating an end to conventional budgets, monthly accounts, variance inquests and cross-charging.

BRANDING AND BRAND POSITIONING

A catchy slogan or technically brilliant advertisement is not branding or brand positioning. Rather, the whole sense and feeling a customer has about a supplier, both directly on the senses or almost subliminally, form the brand both corporately and within a segment, division or product and service.

The brand is the very 'being' of the business, what it stands for in the customers' minds and even in the community or nation. As such, every employee needs to live the brand and every internal process has to deliver it. Customers will brand a business whether the company has specifically branded itself or not. Branding and brand positioning cannot be left to chance. It requires some science to achieve the desired result in the marketplace and is a key driver of changes to internal capability, both in processes and behaviour.

The definition of a brand

The minimum definition of a brand is that it is a means of naming, identifying, characterising and describing a product or service. It is the process of naming differential qualities in order to own them in the users' minds. A brand authenticates and guarantees the quality and performance delivery of what is being offered. It is a mark of recognition and reassurance. And in this minimal sense, branding is at least as old as the silversmith's hallmark.

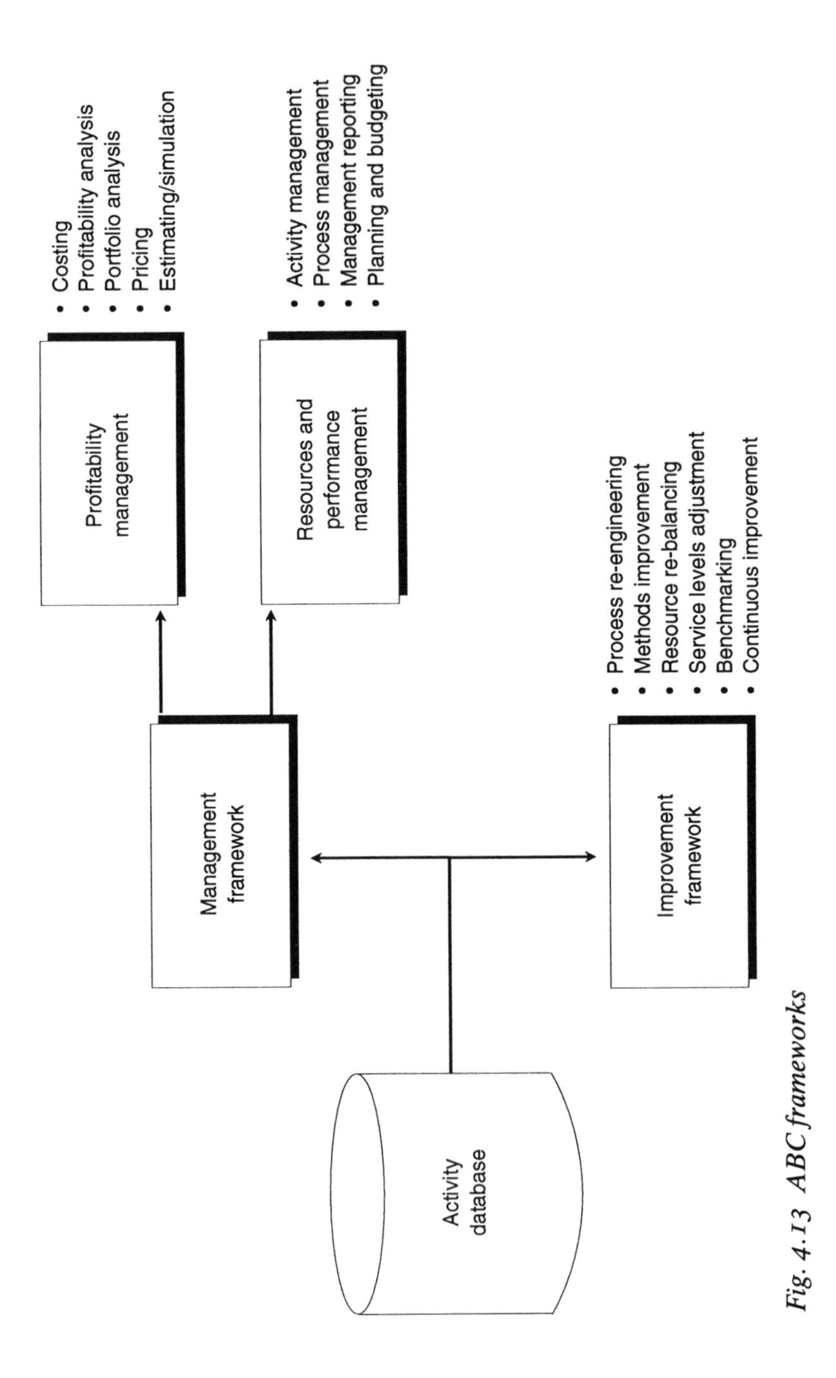

Fig. 4.13 ABC frameworks

Fig. 4.13 ABC frameworks

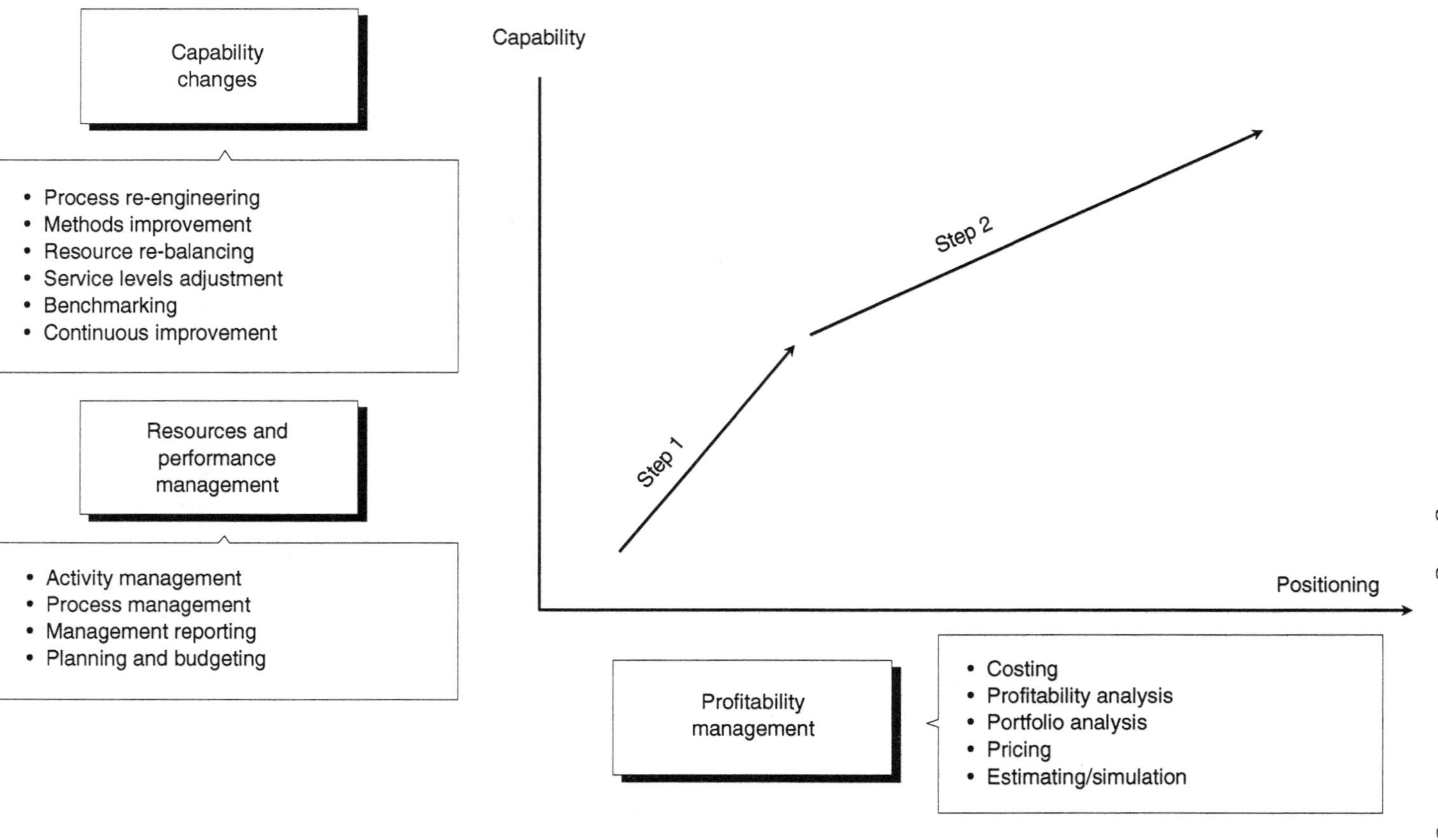

Fig. 4.14 Making the inextricable link visible

To be credible, the process of creating a brand needs to be much more than selecting some desirable slice of psychology and bolting it on to a product or service offering. The product's or service's function, performance and characteristics must link or integrate relevantly and credibly to the fundamental human needs that the brand wishes to represent. Societal trends and the initiatives of competitors influence customer preferences such that the appeal of the brand and the desires of the customer, are subject to evolution and change. Some brands are fortunate to be able to sustain their life through significant market changes over many years.

Brands as symbols

The strongest, most successful and enduring brands contain values and imagery which reach far beyond the narrow bounds of functional utility. They are anchored in human concerns, emotions and aspirations. Such brands are symbols – complex, densely packed sets of values, capable of evoking and conveying a variety of facts information and motivations. Fully developed, brands are composites, syntheses, of both rational and non-rational components, prompting sympathetic and favourable responses in their target audiences. Shopping at certain supermarkets means more to loyal customers than just good quality food at low prices. Certain branded shoes or trainers do more for wearers than just keeping their feet dry or protecting their feet while jogging.

Big brands succeed by employing universal human motivations. They manage to associate themselves with their users' own hopes, fears and motivations in such a way that they become complex symbols of them. The spectrum of these motivations is enormous. Success derives from relating a selected cluster of psychological values credibly to the functional purpose of the brand, so that these values illuminate and enhance the brand's physical function. By offering a discrete set of coherent functional and psychological needs and values a company can create distinctiveness and differentiation, provide a clearer choice to customers, win customer preference, and therefore create added value and profit.

Brands as assets

These positive associations of a brand are its most valuable property. The brand owns its image in consumers' minds. Brand images planted ineradicably in the minds of users become a company asset, intangible capital which does not appear on company balance sheets.

Advertising and other forms of support activity which sustain brand imagery should reasonably be considered an investment rather than expenditure, as investment should be judged on long-term as well as short-term payback. For many companies, the brand is at the centre, the heart of activities.

The definition of brand positioning

Brand positioning is the process of owning in consumers' minds the most motivating needs, both psychological and functional, in a market – in other words, the external positioning as perceived by the customers. It is what the organisation stands for and evokes in the customers' minds, rather than purely the products or services it offers.

Corporate brand positioning is the development and ownership of a clear and competitive identity that can be translated and delivered by the organisation's internal processes, internal values and beliefs (its culture), and the internal structural framework – in other words, the key elements that constitute the organisation's capability.

Brand positioning is an integral part of the BPM journey in that the internal capability of an organisation, in all its facets, must be aligned with the external positioning the organisation has chosen in order to generate revenue. Any conflict between positioning and capability weakens the perception of the products' or services' value in the customers' minds and reduces revenue, while increasing internal organisational and process stress which increases costs.

The management task

Ensuring that the company brand is optimally positioned for the future is therefore a vital task for management. Yet it is not an easy one to fulfil. It is particularly complex in the case of service brands, where the whole brand experience derives from people interactions. It cannot be left to the marketing department. It has to embrace, and be embraced by, the whole organisation. People have to live the brand.

Issues facing the service sectors

With time and patience, a concentrated period of watching television advertisements will highlight the differences between product advertisements and those for services. Product advertisements, by comparison, provide clearer messages and evoke stronger responses through better matching of the product to the target customer. Services advertisements have a tendency to confuse and often bewilder the customer as no clear psychological need is being met and the functional elements of the service are generally in an undifferentiated market.

However, when economies have been depressed, the service sector has difficulty meeting any upturn, particularly where the market size remains smaller and more service providers have entered the market. Differentiation based on price alone leads to a shake-out among the suppliers. The service sector, relative to product brand positioning, thus faces the greater challenge in achieving brand positioning and supporting this through radical and innovatory changes to their capability.

The definition of branding raises four key considerations for brand-building in service businesses:

1. Deciding if a brand potentially exists separate from the 'corporation'.
2. The relationship of a brand to the corporate parent.
2. Persuading operations departments that segmentation is worthwhile and possible.
2. Defining what staff have to contribute.

Issue 1: Can a brand exist in the service sector?

Most service businesses are still at the pure functional stage of market segmentation. Banks look at consumer groups by age, social class, salary, number of cheques they write, size of overdraft. Airlines look at frequency of travel, fare paid, destination point. Hotels look at location, facilities offered, number of stars, and so on. It is this kind of analysis which gets us to special savings accounts, 'budget' fares, 18–35 travel clubs, chains of restaurants, and so on. There is nothing mistaken with any of these 'slices' but they are 'functional' slices which do not embrace a relevant motivating customer psychology. There is also a sense in which demographic and operational segmentation will lead us up some blind alleys. These may lead to cost effectiveness from the supplier's point of view, but are insensitive from the point of view of consumer psychology.

Nobody likes finding out that the person making the same journey in the next seat is paying half the price you paid yourself. If you are in a hurry to pay for your petrol it is not immediately clear why you should be in the same queue as chocolate bar purchasers and video hirers. The description of the Gatwick North Terminal as a supermarket with parking places for aeroplanes is not far from wrong, and why does an expensive country-house hotel weekend break get mixed up with a large company's sales conference?

As businesspeople we can all clearly understand what is going on in these different situations. We know that the deals are to ensure that otherwise spare capacity is being used. We are being targeted as we recognise certain segments we occupy. Spin-off sales are being created. However, psychological wires are getting twisted as people with different needs and motivations are getting unwillingly lumped together.

Genuine brands have to be a coherent offering to a cohesive target group with a consistent standard that does not get pulled apart by the pressures of the short-term operational efficiencies and short-term measures.

The key to clear positioning is in the psychology. Deciding if there are viable 'brand' slices in a market must begin by seeing if there are discrete clusters of different psychological needs or motivations. The 'needs' map we need to create of a market must reach beyond the conventional demographic segments as it must examine relevant attitudes and motivations. If we are in travel, what role is travel playing in our customers' lives? Is it relaxation, stimulation, celebration or is it a search for freedom? Is it a support system for a busy successful working life? Is it an expression of status and importance? Or is it a way of forging relationships? It is rarely an end in itself . . . so what is the purpose . . . the real purpose? Self-improvement, self-expression, self-indulgence, self-advancement?

The same sorts of issue surround hotels and restaurants but often they are reduced to categories like business and leisure. Hotels and airlines know that businesspeople are forever having 'meetings', but they do not seem to be interested in what the meetings are for. There are lots of different kinds of 'meetings' with different objectives, different needs and different constraints. A closer fit to the real needs would provide clearer differentiation and lead to customer advocacy. In the undifferentiated market, convenience is often the only choice the customer can make, while knowing that the meeting room will generally be unsuited to the specific purpose of the meeting.

In the financial sector, the role which money actually plays in people's lives has to be understood, and the attitudes which different kinds of people have to it. Money is only rarely an end in itself. What do people want to do with it? Does it control them, or do they manage it? Does it represent security or anxiety? Do they spend it or save it? Do they flaunt it or feel guilty about it? Is it all secret or is it a status symbol? Banks, building societies, and insurance and pension companies have competed consistently on price in a declining market. Employee motivation to sell, driven by commission payments, leads to unsavoury selling practices and poor public relations for the whole sector.

In the service station field, what exactly is going on in a petrol forecourt? Are these distress purchases interrupting the progress of busy people who want simply to be on their way, or are they 'welcome breaks', in every sense, from the madness and competitive nature of the motorway – sorts of 'oases' of calm. Examples of creating differentiating positioning in the petrol retailing sector are explored in more detail later.

In clothes retailing, the role fashion plays in individuals' lives has to be understood. Is it exhibition, adventure, conquest, group security, status, self-expression and what conservative taste has to say about somebody's motivations? Home improvement, equally, embraces a wide span of consumer psychology. The differences in customer motivations for purchase are the raw material for building brands.

What is fascinating about even such a cursory overview of the services field is that the range of potential psychological appeals available is not only wide but is also profound and real. In contrast, positioning a portfolio of confectionery, cigarettes, washing powder or snack products is often about 'force feeding' or even inventing psychological appeals. There is no true sense in which certain cigarette brands are about 'freedom', or a washing powder about 'mother love' or thin mint chocolates about 'status'. But services impinge on the real, central, driving forces of life; money, health, self-fulfilment, control, progress, adventure, security, job success. In the service sector, you don't have to invent anything.

How is it then that while most competitive packaged-goods markets have been segmented all ways round in non-functional terms, so that your choice of whisky, beer, tea, biscuit, chocolate bar – even pet food – carries a raft of psychological overtones, so little has yet been explored in positioning services? Indeed, the current trend seems more about developing marginal 'systems' advantages, technological points of difference, all of which are easily imitated in functional terms. Card cash machines, self-ticketing, automatic check out in hotel, inter-active information systems, automatic funds transfers, and so on, are about how customers access a product or service, and although desirable from a customer view-point, they are not the strongest base on which to create a specific brand.

The needs map must examine a market from psychological first principles and see which human needs are 'in play' and which might potentially be brought into play, and which then can be coherently clustered and linked with some meaningful aspect of the functional mix. For example, if a brand in travel is to be built on status, fast service and separation, can we actually put together a functional product to support these appeal areas? If our service stations are to be an oasis of rest and calm, can they deliver?

Unfortunately, even if a clear brand position has been defined, the packaged-goods marketing manager arriving in a services environment quickly realises that you cannot undo fifty years of cultural and organisation history overnight. Typical expressions of the inertia can be found everywhere:

- 'Sorry', says the building society chief executive, 'you can't have separate branches for savers as opposed to borrowers.'
- 'No', says the airline chief, 'we can't separate business and leisure travellers in the economy cabin.'
- 'I'm afraid', says the oil company retail chief, 'the same service station will have to do for the quick in and out motorist and the leisurely family shopper.'

Against these kind of segmentation odds the solution seems too often to give the operators merely a 'taste' of branding. Let's cluster what we can and give it a name.

The result is that companies slice 'bits' out of the universe and brand what's 'easy' to brand. Interestingly, the first-to-be-branded segments tend to be the premium offerings of the more prestigious services. This raises two serious problems. First, trying to give a specific segment a stand-alone quality ignores the fact that a significant proportion of the product specification is generic and relies on non-branded or corporate factors. If the train or plane is delayed, or the queues in the building society branch are long or the staff unfriendly, or the hotel badly sited, these may not be relevant to your 'specific' brand claims, but they will form part of an adverse customer evaluation. The brand is marked down as a result of failings in its generic corporate parent.

Service 'brands' are usually inextricably part of some bigger whole, or they represent 'added value' bundles which are reliant on the delivery of some basic attributes. If these are not delivered the added value is irrelevant. People do not say, the airline was late but the service and recognition I was given in Club Class were excellent, or the bank branch is miles away and the staff are rude but the interest on my special savings account is very good.

The second problem is that the more successful you are at branding the 'premium' products, the more the corporation, as opposed to its brands, becomes associated with the unbrandable, commodity end of its product range. The railways could come to mean just crowded commuter trains. An airline could just mean cheap economy fares, the bank just one ordinary savings account, and so on. This may well be where the mass turnover is, but its unlikely to be where the prestige or the corporate image aspirations are. This leads us to the important second issue – how the brand-clusters should relate to the overall corporate position.

Issue 2: What is 'Brand' and what is 'Corporate'?

First we need to decide on our branding model. Borrowing again from package goods there are really three basic models:

Model 1 – Standalone brands;
Model 2 – Corporate name + functional descriptors;
Model 3 – Corporate + 'sub-brands' where both the corporation and the sub-brands embody real added value.

For most service businesses, Model 1 is simply too expensive. Model 2 is most common and fairly straightforward provided the corporate brand is well defined. Model 3 is, by far and away, the most rewarding route for big service companies, but does need some careful handling.

How can the optimum relationship of individual 'brand' slices be established with the overall corporate position, where the ideal is that brands should 'positively' borrow something relevant and useful from the company image, reputation and positioning. What is absolute folly is for either to detract from the other. In broad terms, it is the 'over arching' corporate position which can best tackle the consumer's need for reliability, accessibility, security and so on, and the brand which can explore specific 'functional' slices.

The corporate position has to be worked through as the broadest and most competitive concept available which supports the individual brand clusters. The corporation or parent should represent a positive set of 'core' values, evidenced by some measurable aspect of the operation. Friendliness, approachability or care are simply not good enough. The corporate values must genuinely apply to all the brands, or sub-brands. They then represent a solid platform on which 'premium' or 'targeted' brand offerings can be built, so that a brand then becomes quite clearly

Corporate core values + Specific targeted proposition

It should almost be a golden rule to build the corporate parent name firmly into the sub-brand presentation. In this way, a two-way synergy is developed. Perhaps, more importantly, this discipline ensures consumer coherence. The main challenge is to get a focus on the corporate core values which represent a meaningful customer objective.

So often the slogans seem designed more for the City than the customer. Typical examples are:

BRINGING OUT THE BEST IN THE WORLD
THE NATION'S BUILDING SOCIETY
GERMAN KNOW-HOW IN GLOBAL FINANCE
NOTHING VENTURED, NOTHING GAINED
FOR PEOPLE WHO VALUE THEIR MONEY
LEADERS IN LEISURE

The kinds of area which can provide the basis for a serious corporate position, which are linked with genuine customer needs, are in areas like:

RESPONSIVENESS	(Staff–customer ratios, response times)
SECURITY	(Size, guarantees, service contracts)
LOCAL KNOWLEDGE	(Branch network, local staff)
EXPERTISE	(Staff training, research qualifications)
QUALITY	(Commitment to specifications, suppliers)
INNOVATION	(Leading edge always)
LONGEVITY	(Systems compatibility, servicing availability)
TIME-UTILISATION	(Business-oriented systems, procedures)
CONVENIENCE	(Location, range, choice, delivery systems)

Price, warmth and friendliness, promises to do better and being trusted, are more like hostages to fortune.

Issue 3: Convincing the operations departments

Services are often based on very expensive assets, such as aeroplanes, trains, real estate and computer systems, which have to be optimised and which have traditionally been managed centrally. They are truly fundamental to the efficient operation and profitability of the corporation, in that if they are not properly managed, disaster looms. Efficient management usually means operational systems and expertise of a specialist but undifferentiated kind. People who manage property are expert at property management, people who buy aeroplanes know about load factors and fuel efficiency, people who put millions on overnight deposit know about money markets. Their task and the judgement of their performance is against broad-scale corporate criteria. Segmentation of any sort invariably shows up in worse performance against these criteria.

Operations managers in service companies are very powerful people, and in an absolute sense the company relies on them. They can always produce statistics to prove that the return on assets is optimised by centralised, broadly based activity, and that tailor-made differences will cause havoc with systems, efficiency and profitability. Culturally, they look on customers as the raw material to be processed. Such managers are the masters of portion control, yield per sq. ft., traffic flows, capacity management. They are a hostile environment waiting for any notion of brand positioning and customer perceived differentiation to appear.

But it is precisely these operations people who are absolutely determinant in delivering the basic essentials of any service offering such as price, schedule, punctuality, convenient locations and so on. A crucial balance has to be worked out between the need to think and have processes deliver in a segmented manner and the efficiency drivers associated with central and single focused operations. The operations departments have to understand the case for and the commercial advantages of 'branding', and the marketeers need to appreciate and be aware in detail of the operating inefficiencies which segmentation creates. The company as a whole needs to be capable of deriving how the pluses and minuses net out.

The strains of a recession bear down particularly hard in this area as operations departments are always called on to increase efficiencies. The approaches are well tried and successful in the short term. Branches close, staff ratios are reduced, flights are consolidated, price differentials are widened by bargain sales, service frills are eliminated, special discounts are offered for favoured target groups – again, all understandable when in business survival mode. But so called brands, which were based on cosmetic, marginal differences, will simply be submerged by this sort of fish-market activity. Even in hard times, the service sector businesses must be seen to be dedicated to meeting the customers' real needs rather than ensuring their own short-term profitability.

Issue 4: Implications for staff

One area, in particular, in the 'services branding' field is fundamentally different from fast moving consumer goods or product marketing. In the service sector staff have a key role to play in creating and sustaining the brand.

A vast amount of effort has gone on into 'Customer Care' or 'Putting people first' programmes and so on. The initial focus was aimed primarily at changing the behaviour of operators, shop assistants and other people in regular customer contact. While claiming some success, many of these programmes served only a superficial purpose, many ending by alienating both staff and customers.

Customer expectations were raised, but the processes failed to deliver consistently. Staff knew from their hearts the issues plaguing customers, but lacked the empowerment to influence the processes outside their control that would truly create care for the customer. The issues that staff cannot normally influence are interdisciplinary or they require high-level decisions or are enmeshed in organisation politics or are outside of the comprehension of those in ultimate authority. There is a need to get beyond the smiles into the systems, organisational structures, product specifications and performance standards to ensure that the people up front have:

- a decent product to sell, tailored to customer needs, not the best that can be produced by the existing organisation for which staff have to keep apologising, however charmingly – in other words, a proposition that leads to retention and advocacy;
- systems, procedures, organisation frameworks, appropriate measures and management behaviours that are collectively focused to create the desired differentiation – in other words, the corporate capability.

Much of this focus can tend to default to achieving perceptible improvements in the functional aspects of service offerings. However, if a company is to create 'meaningful' brands, it will be increasingly necessary to examine how staff can contribute to the non-functional or psychological aspects of the mix.

There is a need to create a 'brand mind set' in staff throughout a service organisation, where staff have to be encouraged to live inside the brand rather than outside it. The company must understand that the brand is not just something the staff have a relationship within the function of their jobs. Their jobs help form the image of the brand itself. This is not another call for exhortation or simply training. The beliefs and values, and consequent behaviours of managers, and a change in their role from control to one of helping others has been found to be the factor that can leverage the change inside a business.

The service sector will not be able to avoid the issues concerning brand positioning. Creating a differentiating customer proposition founded on providing a service involves the mind, and an understanding of brand values, resulting in an appropriate quality of service and behaviour.

In fmcg marketing the concept of a brand role is well established. The definition of the brand role is conventionally used in writing advertising briefs, in specifying casting requirements and so on. In the services sector, the staff have a role, and where the service is differentiated by brand, the roles will differ. The role the staff play delivers the customers' needs, and if the needs are different, the role must be too. Service staff increasingly will not only have to be completely familiar with the functional details of the brands they are delivering, but they will also need to understand the motivations and needs of the different customer groups. Staff will need help and new techniques in enacting the required roles, they will need different degrees of empowerment, and they will need the support of the entire organisation through its structures and business processes.

The worrying trend, although many organisations have not seen it yet, is that they are learning the need to profess a service differentiation through advertising, but few have successfully tackled the need to shed decades of inferior and inappropriate capability. Beware the company charter issued to customers and the exhortation of the vision statement issued to staff.

Brands which succeed over time, in a competitive environment, are typically extremely complex properties. They consist of a complicated clustering of many differing constituents, integrated into a coherent whole. The whole is

Basic need	*Which is about . . .*
Achievement	Need to strive, reach for goals, be successful.
Dominance	Need to be or feel in control, of self and others.
Autonomy	Need for space, privacy, independent thought.
Inviolacy	Need to preserve one's self-respect, good name.
Avoidance	Need to avoid stress, pain, danger, anxiety; peace of mind.
Acquisition	Need to accumulate possessions.
Cognisance	Desire for knowledge and understanding.
Order	Desire to make sense of, to structure rational frameworks.
Deference	Need to accept hierarchy; praise, conform to superiors.
Succourance	Need to be cared for, comforted, guided, protected.
Passivity	Desire for rest, easy life, low profile.
Sentience	Desire to experience/feel through senses.
Affiliation	Need to belong; friendship, co-operation.
Nurturance	Wish to care for, look after, protect, be parental.
Recognition	Desire to be recognised, flaunt, be seen, extolled.
Play	Need to have fun, amuse, divert; childish pursuits.
Sex	Desire for erotic/sexual relationships; be in love.

Fig. 4.15 Basic human needs

greater than the sum of the individual parts. Or put in commercial terms: my brand has a bigger share of the market than yours, even though both spend similar market support budgets in similar media and seem to offer, in objective or physical terms, something very similar. My brand has synergies at work within it which make 2+2 add up to 5. Your brand's 2+2 = 4. The difference between my 5 and your 4 represents the much greater profit made by my brand, year in year out.

Achieving success will not be intuitive. Brand positioning to achieve a differentiating customer proposition is a science, a science to adopt on the BPM journey.

Customer needs mapping

A method of establishing brand positioning for the service sector starts with competitive market and consumer needs mapping. Achieving positioning is founded on understanding that:

- human needs are the most meaningful positioning and segmentation differentiators;
- to be convincing, psychological need must credibly derive from and link to the product/service performance;
- a way has to be found to display all product/service relevant fundamental psychological needs, against which brands can be mapped.

This is established by:

- finding a way to review human needs for product/service relevant subsets;
- producing a category-relevant list of needs;
- mapping brand positions and performance against these needs.

Basic human needs

In the 1930s, Murray, a psychologist, successfully catalogued human needs as being to do with such things as inviolacy, autonomy, avoidance, succourance, dominance, and so forth, which lead to descriptions of the product/service relevant subsets. A description of each of these human needs is shown in Figure 4.15.

Taking one these needs, we see that 'achievement' is about the need to strive, reach for goals, win, be successful. A brand can serve aspects of this need. For example, within the overall need for 'achievement', driving a high-performance car could be said to be about:

- making things happen;
- celebrating achievement;
- aspirational fantasies;
- being an adult;
- boosting one's confidence;
- winning against others.

Within these general categories, a subset of needs can be developed which are relevant to driving high-performance cars. For example, 'making things happen' may be about action-orientation, destiny-attainment, and so on. Customer needs and competitor products and services can be mapped against these subsets. Taking a sector such as petrol retailing, a list of relevant subsets of human needs can be developed. Some examples are shown in Figure 4.16. This table would be continued for all the basic needs.

Basic need	*Which is about . . .*	*Relevant subset (RSS)*
Achievement	Need to strive, reach for goals, be successful	1. Control progress 2. Shrewd 3. Want responsibility
Avoidance	Need to avoid stress, pain, danger, anxiety; peace of mind	1. Reassurance, protection 2. Risk aversion, caution
Succourance	Need to be cared for, comforted, guided, protected	1. Known personally 2. Relationship 3. Helpful, simple
continued . . .		

Fig. 4.16 A relevant subset for petrol retailing

Scoring yourself, your competitors and customer needs

For each subset, the sector's ability to deliver on this need can be scored (1 = not at all, 5 = easily), the degree to which the need is a hygiene factor (1 = not at all, 5 = minimum expectation), the actual achievement of the company (1 = not at all, 5 = totally), the advantage over competitors (1 = no advantage, 5 = completely ahead), and the degree to which the company can develop delivery against the subset (1 = impossible task, 5 = can develop with ease).

Scoring should be undertaken using a multi-functional workshop approach of internal experts with knowledge of external customer needs, customer trends, competitor initiatives, and other non-competing sector customer influences. Obtaining this data is outlined in other sections of this chapter. Let us assume that the exercise produced the result shown in Figure 4.17. From the analysis, the company would be looking to gain an advantage over the competition against the relevant subsets, particularly where the sector ability was weak, but the company's development potential to deliver was high.

Basic need	*RSS*	*Sector ability*	*Hygiene factor*	*Actual achieved*	*Advantage*	*Degree develop*
	1	1	1	4	3	5
Achievement	2	3	1	3	4	4
	3	2	1	4	3	5
Avoidance	1	3	2	2	2	2
	2	3	5	1	1	5
	1	3	2	1	1	3
Succourance	2	4	1	3	1	2
	3	4	3	2	2	2

continued . . .

Fig. 4.17 Looking for differentiation

In this example, the overall conclusions were summarised as follows:

- Some needs were process needs, about the process of doing business (the How). These would be categorised as the evaluators.
- Others were user end needs that related to what the company was in business to deliver (the What).

Interestingly, for the petrol retailing sector, the overall scoring analysis did not favour the company attempting to claim delivery in areas of need where only the user could define its scope (e.g. autonomy, acquisition). Put another way, this prevented the company taking the path of positioning itself as being able to deliver every hope and ambition and make every individual and family dream

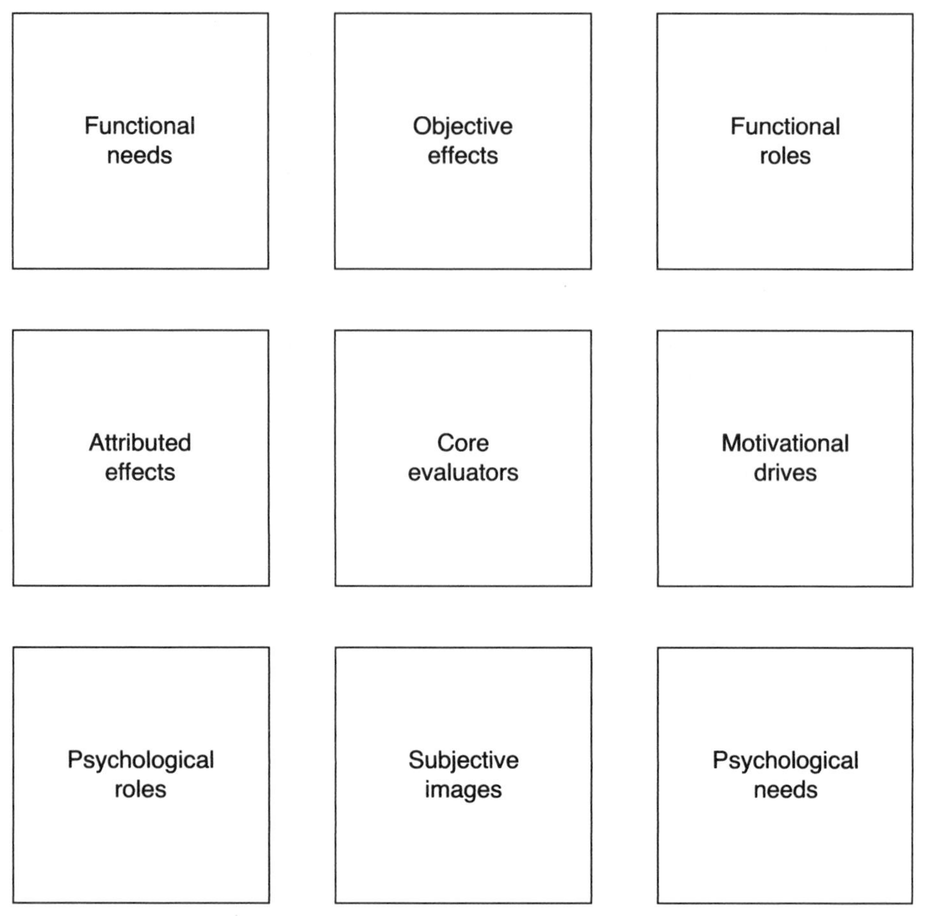

Fig. 4.18 The basic brand positioning matrix

come true, just because they had lead-free petrol. Some companies have claimed such lofty powers through buying one of their products or using their services.

In summary, the analysis showed that in the petrol retailing market, where the basic product, petrol, is essentially a differentiatable generic offering, then service is the only way to offer the consumer any element of choice. Surprisingly, 'succourance' and 'achievement' came through strongly and were used to create two specific brand positioning matrices to appeal to two specific consumer segments.

As we will see later, this can have significant implications in terms of the capability of the organisation, relative to its traditional business processes and culture.

Creating a brand positioning matrix

A number of guidelines apply to nearly all situations:

- customers do not analyse brands;
- everything is part of everything else;
- the brand is a whole, an overall holistic entity;
- each element must be a cohesive part of the whole;
- customers respond to the whole;
- to build brand loyalty, all the parts must be managed;
- customers must be taught to value most what your brand does best.

As outlined earlier, the positioning constituents are functional, psychological and evaluative and can be shown in the form of the matrix in Figure 4.18. The matrix is then used to complete the relevant subsets as key strategic elements in customers' minds.

A matrix based on 'achievement'

In the petrol retailing example to satisfy the 'achievement' segment, shown in Figure 4.19, only some of the subsets are included to retain clarity. The matrix then forms the basis for the persuasion strategy; the proposition that retains customers and makes them advocates of the company and its products/services.

From the matrix, the message would start to form. In this case: 'All petrol retailing sector providers offer much the same by way of products but for the business traveller it is often a real nuisance to have to stop to fill up and then get tangled up in queues of customers. This company has all the products you need, but with a difference. For the busy traveller we try to make it genuinely easier for you to meet your basic need and then get you speedily back to your journey.'

The implications of this positioning then impact on the capability to put in place. The busy traveller is not so interested in the petrol price. In many cases this will be a company expense. They could be more interested in not having to get out of the vehicle and risk their business clothes smelling of petrol. They will be used to having credit/debit cards. In other words, the petrol filling and paying needs to be streamlined and non-invasive and got through as quickly as possible.

One could imagine a differently designed service area that had a cover extended over the filling cap of the vehicle, allowing an out-of-view attendant to fill the vehicle. The driver, in the meantime, selected sweets, cigarettes, newspapers and so on from a panel next to the driver's window. The selected items being brought automatically to the driver and released after payment by swiping a card. After payment a barrier could release the vehicle, such as is found at exits to car parks.

A matrix based on 'succourance'

In the petrol retailing example to satisfy the 'succourance' segment, shown in

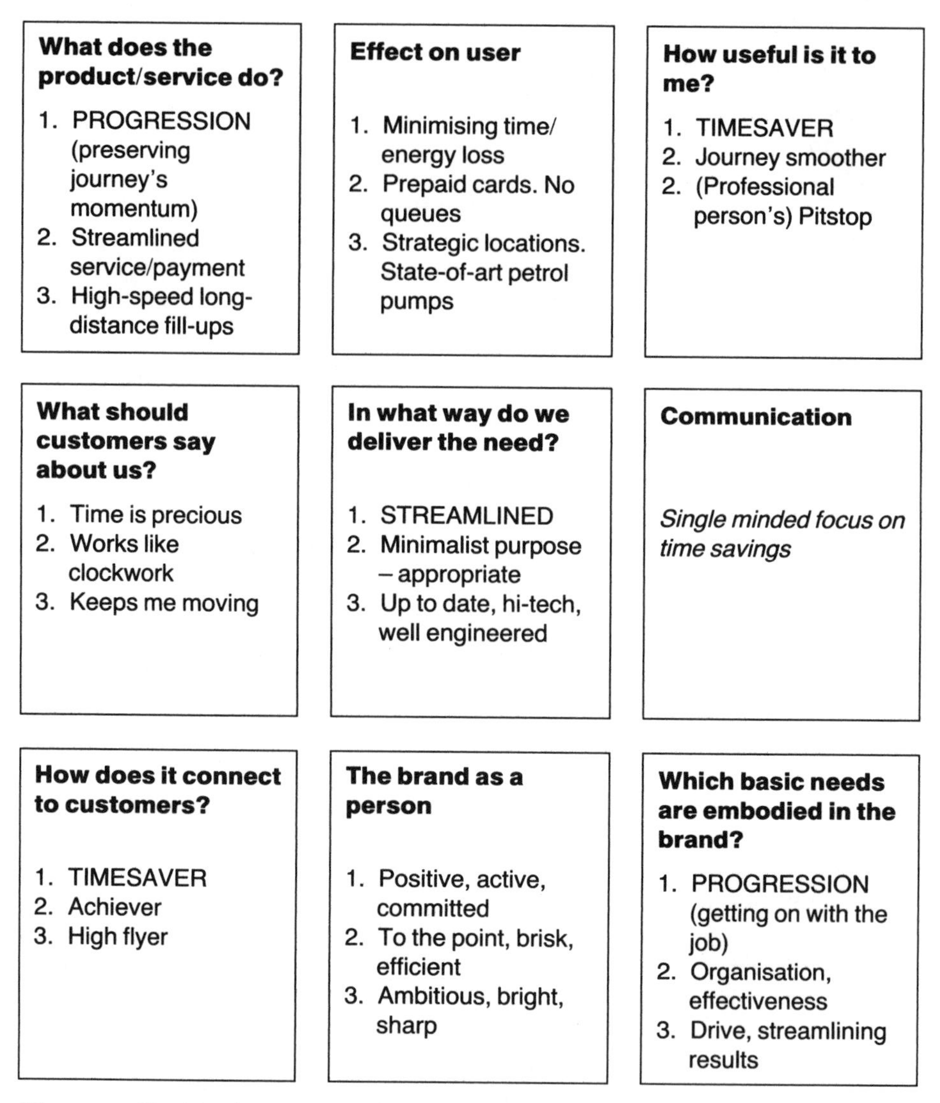

Fig. 4.19 Positioning matrix – 'achievement'

Figure 4.20, only some of the subsets are included to retain clarity. From the matrix, the message would start to form. In this case: 'All petrol retailing sector providers offer much the same by way of products but for the person who sees going out as an expedition that includes filling the vehicle with petrol, we can make the whole experience one of a useful way of spending time, enjoyable in its own right. This company has all the products you need, but with a difference. For the casual purchaser, the service centre provides a one-stop shop for your car and associated household problems.'

What does the product/service do?

1. EXPEDITIONS (weekly fill-up)
2. Full range of car care services
3. One-stop shopping for your car

Effect on user

1. Making the most of (partner's) shopping trips
2. Car wash/fill-up/DIY
3. Suburban forecourts/ supermarket adjuncts

How useful is it to me?

1. TIME FILLER
2. Wash and brush up
2. Resource stretcher

What should customers say about us?

1. Worthwhile use of (otherwise empty) time
2. Keeping my car well looked after
3. They've got everything I need – and it's great value

In what way do we deliver the need?

1. PURPOSEFUL
2. Comprehensive and spacious
3. Discounts, great prices

Communication

Single minded focus on value and choice

How does it connect to customers?

1. TIME FILLER
2. Boredom relief
3. Aladdin's cave

The brand as a person

1. Practical, helpful
2. Easy, relaxed
3. Cheerful, supportive

Which basic needs are embodied in the brand?

1. EXPEDITIONS (enjoyable outings)
2. Diversion, useful occupation
3. Value, choice

Fig. 4.20 Positioning matrix – 'succourance'

The implications of this positioning then impact on the capability to put in place. This segment is more interested in the petrol price, is likely to be making more of a statement with the vehicle, wants to (has to) undertake more DIY, has time to kill, and is looking for all round good value.

One could imagine the service area being just that – all about service. Not an air gauge out in the rain or the car wash stuck round the back. The whole centre would be geared to give the basic needs plus advice and other DIY facilities. More of a club atmosphere where people with like needs can congregate. If such a facility were close to a main shopping centre then the driver could drop off the partner to tour the shops while the driver attends the 'garage-club'. The shopping load would be brought to the 'club' and loaded into the vehicle. A relaxing coffee completes the experience.

Impact on an organisation

The type of focus that the positioning analysis creates can easily be undone by a company attempting to increase revenue at the margin, by introducing an increased range of products or services, where some of them run counter to the overarching theme. On the other hand, removing a marginal service on the grounds of cost-cutting could easily undo a significant part of the customer relationship the overarching theme had created.

Also, customer segmentation may indicate that although the positioning is powerful and generic, the specific delivery may need to respect the various customer segments. We may all be driven by the same basic human needs in relation to the company providing the service, but our experience of a common proceduralised delivery process may still leave some customers confused, and at the other end of the scale, some customers may feel they are being patronised. As mentioned earlier, procedures put in place in the interests of standardisation and efficiency can still undo the brand positioning that attracted the customers in the first place.

The implications of brand positioning can have a profound effect on the nature and degree of empowerment that staff have to deliver the proposition. The competences of staff may be out of line with the future need to create a different relationship with customers. The types of facility and their location may need to be quite different to deliver the proposition in a way that is consistent with positioning. The level and type of support from information technology could be quite different from that which just handled transactions efficiently.

Many sectors tend to be operating in an undifferentiated market. However, the 1990s and beyond suggest that a quite different business scenario will exist, compared to what has gone before. In an oversupplied and for some, declining market, brand positioning could well prove to be one of the key challenges, both in terms of getting positioning right, but more so in aligning the capability of the organisation to deliver it.

IMAGE AND ATTITUDINAL TRACKING

Positioning the company, its products and services requires an understanding of the major influences on customers within society as a whole and the way that this changes the perceptions that customers then hold. Products and services that may have done well in one decade can fall outside of customer acceptance as values and attitudes change. In some cases this will simply need a change in, say, packaging design, but for others there could be major changes in the overall capability of the business, particularly the nature of the sales process and changes in the way the customer relationship needs to be developed.

To obtain this understanding requires qualitative research based on questionnaires and extended group discussions across a range of demographic social groupings, life-stage groups, and geographic regions. The research needs to look for evidence of changes in customer attitudes, find the causes of these changes, and ascertain if the changes are influencing the nature of the purchasing decisions customers will make in the future. Against this knowledge, the current positioning of the business will need to be matched to establish if any discontinuities now exist or whether any new opportunities for competitive differentiation can be created.

The findings and conclusions that are contained in this section are those which were valid in the mid-1990s and provide the background for companies to use as a guideline for changing their capability through the next five to ten years. As time passes, customer attitudes will change so the output from any research will need to be refreshed in order to re-adjust the company's positioning.

Identification of changes in consumer values

Over a number of decades, distinct characteristics can be established that have influenced how consumers have perceived the world around them. Some of these are highlighted in Figure 4.21.

Any particular decade will be influenced by the decade that precedes it. In this case, the 1980s have influenced the trends in the 1990s. During the 1980s people summarised how their world got better in terms of being more financially secure, having a better lifestyle, higher earnings, more freedom and choice, better homes and holidays. The 1980s also had their darker side expressed as world uncertainty, third world crisis, environmental issues, rising unemployment, the emerging recession, higher interest rates, concerns over health and education standards.

A significant issue that emerged in the late 1980s concerned the environment coupled with a more cynical view of companies' exploitation of 'green' issues and an awareness that consumer action can change company behaviour. Consumers also carried forward some guilt over the unremitting materialism of the 1980s.

The recession has had the effect of undermining confidence in basic living standards being retained, but the price of the caring 1990s seems too high and too direct a threat to our now fragile living standards. Although there has been identified a widespread rejection of the 1980s values, there still exists a basic

Date	Macro-situation	Values
1960	Low unemployment Low inflation	*Positive* • Youth culture • High spending • Rejection of 1950s values
1970	Rising unemployment Rising inflation Economy faltering	*Confused* • Less disposable income • Greater conformity • Lack of direction
1980	Recession Economic recovery Stock market crash	*Positive* • Opportunism • Materialism • Self-gratification
1990	Rising unemployment City panic Recession	Disaffection with 1980s values
2000		Uncertainty

Fig. 4.21 Changes in value

underlying materialism that creates a guilt in relation to the feeling of helplessness when faced with the global issues that impose on our daily lives.

Impact on consumer behaviour through changes in values

The 1990s will see a more knowing, responsible and more cautiously materialistic consumer who is less open to hype and more open to tangible promises, both material and moral. Consumers will increasingly act as individuals with unique needs, as seen from their perspective. They will want products and services that offer true 'quality', where quality will be defined beyond the normal bounds of the specific product or service and will include environmental and social issues as part of the ingredients of choice. They will be prepared to pay for added value where this comes from the value perceived by the whole relationship with the supplier. The added value can come from the level of information obtained, the honesty and integrity of communications from the supplier, the suppliers links to other 'good causes', and so on.

The apparent dilemma of materialism and guilt can be assuaged by companies that make or allow the consumer to do their bit. Products which meet functional needs and help the environment, or make indulgence guilt-free, or provide a benefit for others, will all have a competitive differentiation.

Societal trends

As with consumer values, societal trends in one decade are the foundation of the trends for the next decade. Some of the trends in the 1980s can now be summarised in retrospect. Demographic and economic changes were associated with more and smaller households, an ageing population, increased ethnic diversity and, as mentioned above, the 1980s materialism. Concerns over nutrition changed from minimal knowledge in the late 1970s, through an over reaction to health concerns in the early 1980s, to achieving a greater balance between nutrition and taste in the late 1980s and a focus on ingredients, origins and food safety. Environmental concerns focused on ingredients, packaging, manufacturing processes, waste and disposal and have all contributed to the emergence of the 'green consumer'.

Taking just one factor, the damage to the ozone layer by CFCs, shows how a concern can take on global proportions and lead to global action with large implications back to companies. The trends for the 1990s and beyond should therefore not be ignored as they will influence a significant proportion of the population. Interestingly, the trends that have been identified for the 1990s are focused on individuals and their relationship with the world around them and less on such things as the environment as a key global focus. The key societal trends are:

- cocooning;
- too many lives;
- customer crusader;
- the world-aware child;
- hyper-individualism;
- life quest.

Cocooning

Some of the evidence for this trend has come from a growth in home and vehicle security systems, the development of sophisticated home entertainment products and services, the decline in restaurant business and growth in quality convenience foods and microwave products. The home is becoming more a place of sanctuary in a dangerous and threatening world. Staying home relates to the avoidance of stress, the creation of a personal comfort zone, and a return to traditional values. This trend will change how people view a home. It will also change where people will want to do more of their buying, as evidenced by the growth in home shopping.

Too many lives

Everyone is busy and getting busier. Individuals have to juggle more roles than ever before and everyone is making demands on your time. By habit now, video recorders are used to time-shift entertainment, with much that is recorded still never watched. Also, when buying toothpaste features as more important and interesting than devoting time to buying many other supplier products or services, such as life insurance, then what a company has to offer is not only a function of its products but also the way it provides access to its products. A significant determinant of customer choice will be the time they have available to deal with a company. The type and location of 'access points' will become a competitive differentiator.

Customer crusader

The aware customer making informed and rational purchases is a threat to many companies and an opportunity to others. Not only will they be able to make better purchases, but they will be much more aware of being ripped-off. The customers' experiences of the 1980s have made customers much more wary. Customers will not assume that companies give advice you can trust, not even where the customer is protected by legislation. Providing adequate and relevant information about corporate policies and behaviour will be a competitive differentiator. Having a track record of proper behaviour, rather than an expression of it through publicity, will differentiate a company from others. The 'corporate brand' becomes a potentially powerful source of competitive differentiation.

The world-aware child

The younger generation, the customers of tomorrow, are increasingly sophisticated and world-aware. They have strong feelings about the environment and world problems. They are brand and design conscience and have a strong influence on consumption through 'pester power'. Also, this group is less intimidated by the use of technology. In the future, technology will present wider opportunities in terms of 'access points' which in the past would have found a lower acceptance or even represented a barrier.

Hyper-individualism

Evidence for this trend comes from a proliferation of niche targeted magazines, customised cosmetics, a growth in adult-targeted products, a growth in the promotion of adult games and pursuits, and an explosion of consumer choices in all categories of need. There is a recognition of individualism and personal differentiation. People no longer want to be account numbers or part of a marketing segment. We now face the spectre of a segment of 'one' in a customer base of millions. Treating customers as individuals has implications in terms of the customer relationship, the knowledge one has of the customer and the way the customer will determine how they would want to access a company for its products or services.

Life quest

The evidence for this trend comes from a growth in self-diagnostics for preventive medicine, high interest in anti-ageing products, development of 'engineered for health' ingredients, and mainstream acceptance of major food issues such as fibre diets, knowledge of the effects of caffeine, salt, fat, etc. The trend finds expression through consumers searching for a better, happier, healthier and longer life. There is a determination to avoid the effects of ageing, taking responsibility for health maintenance, and a recognition of the importance of nutrition in the way we choose to live our lives. Everyone at any age is ageing and society is becoming less age-conscious.

Impact on positioning

The heightened awareness of consumers, both in making value judgements and through societal trends, will mean a greater need for external orientation of businesses and faster response to emerging changes in attitudes, global issues and legislation. Similarly, there will be increased accountability to customers in terms of providing information, being accessible, dealing with complaints, and providing proper service. A company will need to encompass the consumers' interest in the product or service itself, but will also need to encompass the consumers' greater interest in the company as a whole. This trend will raise the leverage that corporate reputation will have on influencing consumer choice.

Consumer uncertainty and changing trends creates opportunities for new products and services which provide clear differentiation. There will be a need to consider the whole business process from cradle to grave, involving the raw material suppliers through to the impact on the environment through acquisition and use of the product or service.

What will not work in the future is special offer price discounting, reducing tangible quality to reduce prices, or cutting back on support values such as customer service. There will be a shift to purchasing 'value', where the criteria that make one product or service valuable will be different for another, but generically, 'value' will need to be real additional benefits to customers. It is in this area that positioning will impact on the company's capability; its own values, attitudes and beliefs, and the systems and processes that deliver the end 'quality' to the customer.

In summary, a company will need to understand the forces that drive customer behaviour if it is to develop a product or service that creates clear differentiation that leads to customer retention and advocacy. Gaining this knowledge is likely to have profound effects on internal capability and can challenge many of the accepted norms of running the business, even though these norms have served the business well in the past. To ignore these major trends in value judgements and societal influences is to risk being left in a back water for a number of decades.

DISTRIBUTION AND CHANNEL MODELLING

Many organisations have enthusiastically embraced business process re-engineering with respect to the internal management processes. However, too little attention has been paid to the way in which organisations actually interact with their market. Such an ability derives directly from the analysis of markets, not at the aggregate level, but in the rich detail that they manifest at a local geographical level. Through building geographical models and reproducing the complex behaviour of groups of consumers it is possible to manipulate the market by delivering the best service to the right customers, in the right place, at the right time.

Companies need to focus on the external market processes which drive the performance of services providers, in particular, the key drivers of market share and customer service at a local level. By re-focusing the business on levels of customer access and by paying close attention to the way in which the business interacts with the customer, performance improvement can be achieved through enhanced understanding of customer needs and the instigation of long-term competitive advantage. The internal capability changes in BPM are centred around reducing cost through improved efficiency and thus increased margins, whereas the external capability changes are focused on increasing market share through direct impacts upon the accessibility of the organisation to existing and prospective customers.

The importance of understanding and planning distribution channels cannot be understated. In the financial services sector, for example, the most successful ability to attract new accounts is generated through branches and an ongoing relationship (transactions, etc) still remains dominated by the branch and ATM network. On the one hand, investment in the distribution channels, in the form of refurbishment, network expansion or rationalisation, is an expensive business. On the other, marginal improvement in the performance of each branch in the network can mean significant overall performance and profit gain for the company. In all analyses of market share and distribution channels, outlet location is the strongest explanatory variable for variations in local market share.

From the analyses and conclusions arising out of a number of real-world model building exercises within a number of major UK, European and US companies, both in financial and other retail consumer markets, a number of common observations have been drawn about the way markets behave. Although much of this is intuitively obvious with regard to one's own behaviour, unfortunately, it is all too often overlooked in the way in which organisations market themselves. As an example, let us consider a major clearing bank with a share of the UK market of, say, twelve per cent. At a postal area level this varies from less than eight per cent in many areas to over seventeen per cent in others. A postal area is the first two letters of a postal code (e.g. 'PE' for Peterborough in the postal code PE19 3NA). There are 120 postal areas in the UK.

However, in Figure 4.22, the postal area where it is weakest, with only seven per cent share, at a postal sector level, this varies from over fifteen per cent to less than three per cent. A postal sector is represented by the numbers following the area code (e.g. 'PE19' for the St Neots sector of the Peterborough area). There are 9000 postal sectors in the UK. The correlation between access points, in this case branch outlets, and market share begins to appear.

Typically it would be rare for such an organisation to consider anything other than its share of the national market. Certainly, few organisations look at anything below TV region level and yet it is at the local level, the space in which potential customers live and work, that the battle for market share needs to be fought.

The key now lies in the process of building computer models to simulate customer behaviour within regional markets, which challenge the traditional approach to marketing services. The approach to developing distribution strategy needs to include the concept of 'the network effect', and should emphasise the imperative of focusing on local markets and empowering local management in order to achieve sustainable gains in market share.

Building models of local markets

Given the importance of understanding the relationship between the geographically differentiated patterns of product demand and the way this translates into business in centres and ultimately in branches, it is crucial to develop intelligence about local markets. The approach to generating this intelligence is to apply model building in order to understand the process by which consumers interact with organisations through a set of delivery channels. The core of the analysis is the modelling of the market through the interaction of supply and demand – what is termed 'spatial interaction models'. This approach differs from some of the more typical types of modelling exercise in that rather than attempting to reproduce a small set of observations (for instance, business levels at branch locations by reference to a number of independent variables), it attempts to reproduce in detail the actual behaviour of groups of consumers within the market. Using such an approach it is possible to focus on what is actually happening in terms of processes rather than the relationship between variables in a database.

The geographical modelling process

Spatial interaction models are underpinned by a number of assumptions based on substantial experimental evidence. In simple terms, the choice of destination that a consumer makes is related to:

- the relative accessibility of that destination *vis-à-vis* the accessibility of all other destinations;
- the relative attraction of that destination *vis-à-vis* the attraction of all other destinations.

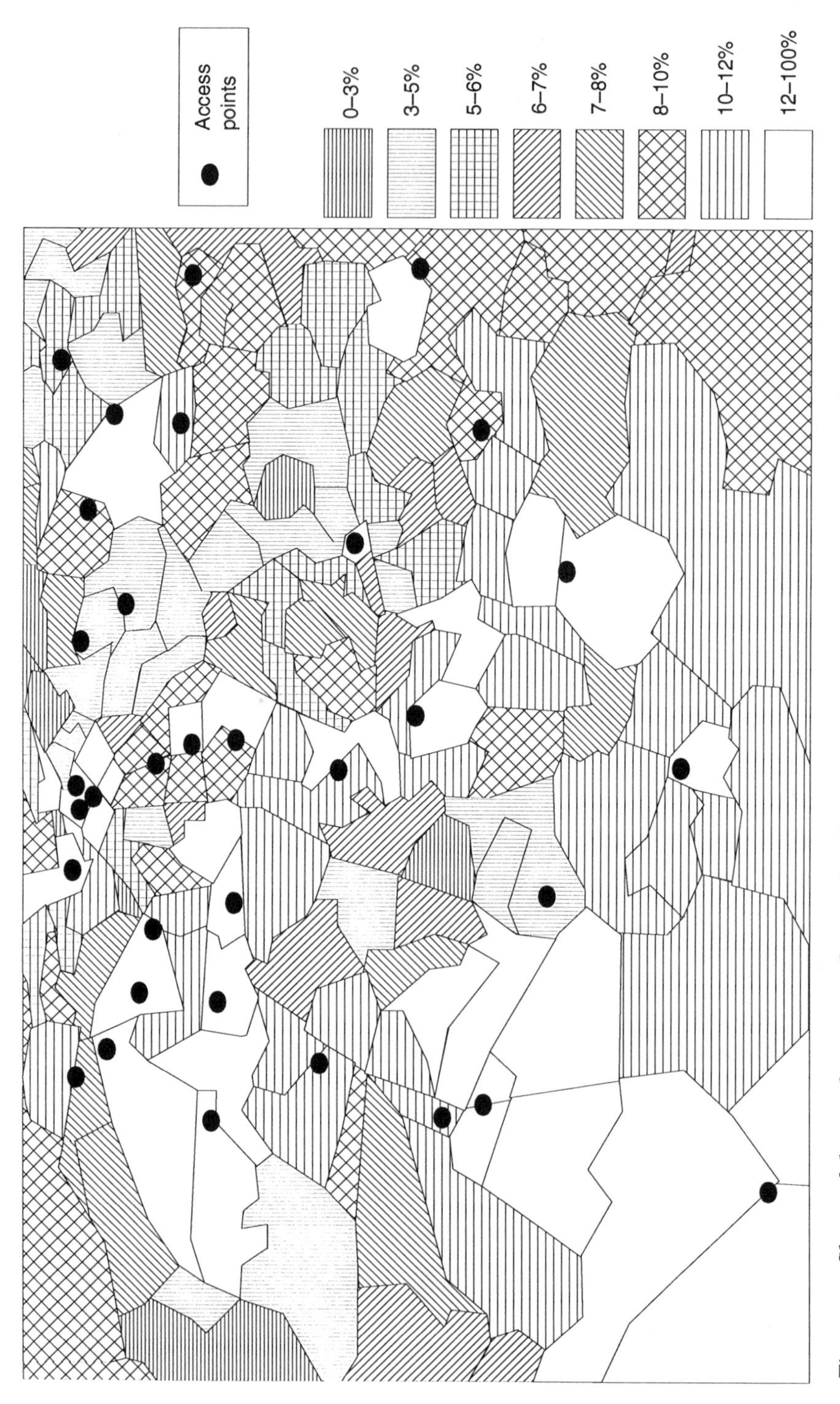

Fig. 4.22 Share of the market at postal sector level

In practice, a much wider set of factors are introduced into the model, such as outlet performance, agglomeration and network effects. The relative importance of all these factors is quantified through a model calibration process.

In some sectors, such as the personal financial services market, the structure is more complex than most other retail markets. In financial services the business is split between recruiting new customers for a variety of products, and servicing existing customers. The recruitment process represents a relatively small number of customers in relation to the servicing of the existing account base, but recruitment is crucial to maintaining and growing the business in the medium and long term.

A simplified framework that forms the basis of the modelling approach is shown in Figure 4.23. There is a focus on two processes – customer acquisition and customer service. Market share is driven by the allocation of customers within the local market between the delivery channel locations of competing companies. Newly recruited customers are added to the customer base of the organisation over time. The organisation then provides service to the customer base through the various service outlets, principally by providing transaction facilities. The provision of good customer service through a network of service outlets is critical to maintaining customer loyalty and minimising account terminations.

In general terms, the modelling-building process is developed in a number of stages. The first stage is to quantify the demand for a range of products within the marketplace. To do this requires focusing on small residential areas such as postal sectors. The postal sector is a small spatial unit for which census and lifestyle data is available on the number and characteristics of the residential and workplace populations. By combining data on known population characteristics with published market research data on propensities of different population segments to consume certain products and services, robust estimates can be derived, for example, of the number of new accounts of different types being opened by the residents of a small area. By focusing on a set of core products this provides an accurate measure of recruitment, which will be spread across different services providers according to their market share. Each new account opened represents, or at least correlates to, the number of new customer relationships being initiated inside a residential area.

The second stage in the modelling process is to quantify in some way patterns of supply within the market. Supply points are concentrated at particular discrete locations, which is in contrast to the origin of demand which is spread, somewhat unevenly, across residential areas. For the financial services sector, outlets or branches belonging to different organisations are clustered to form financial services centres. In the UK there are approximately 2,800 identifiable clusters ranging from major city centres to suburban service centres and small rural towns. In total these account for the 18,000 financial service outlets.

While at an aggregate level, such as a TV region, supply and demand are generally in equilibrium with only slight regional variation, at the local level the patterns of supply and demand are not at all well matched. Therefore, in order

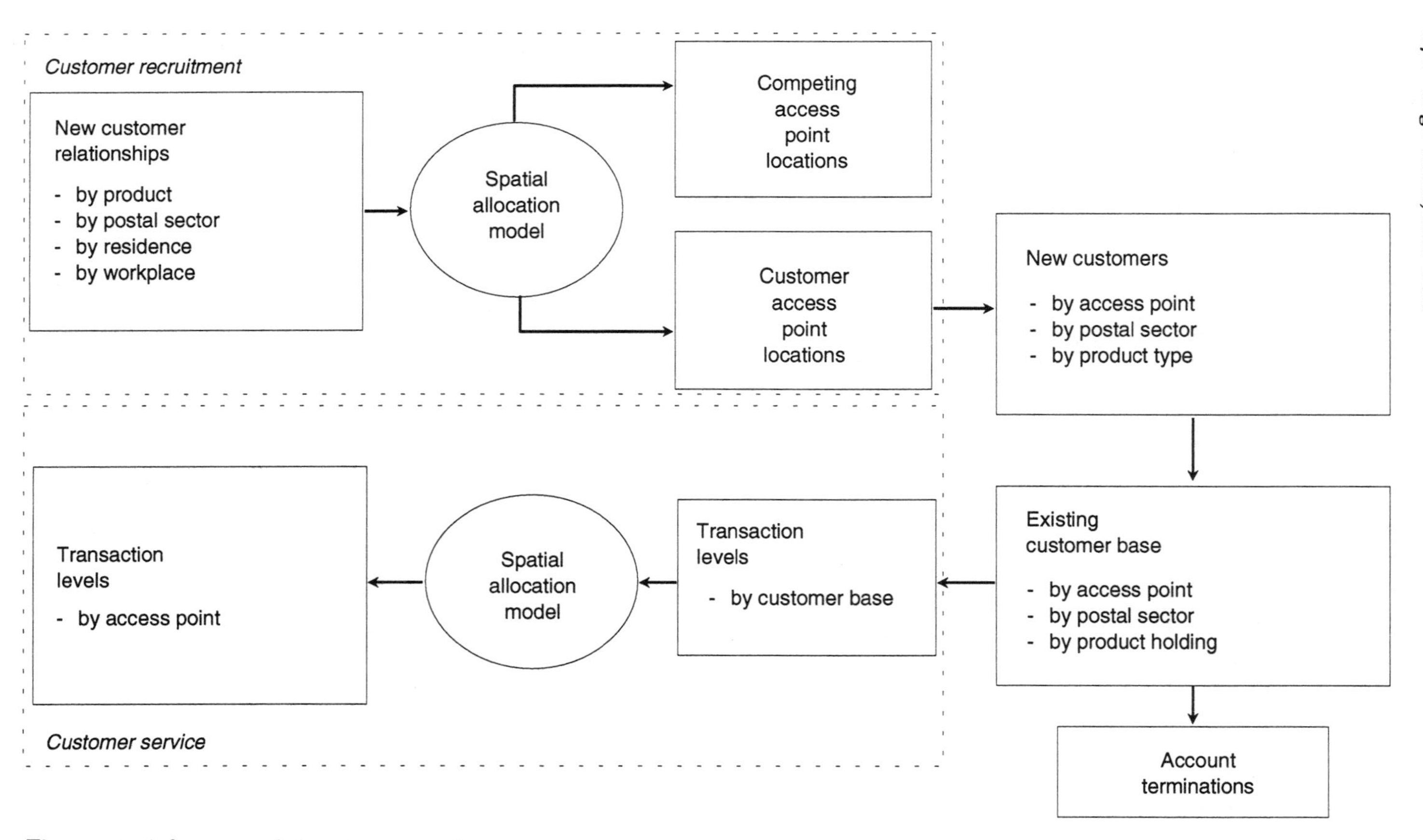

Fig. 4.23 A framework for the external business process

to understand the patterns which arise it is necessary to reproduce, using a model, the process by which all the demand, quantified at the small residential area level, is met by the supply quantified at the financial services centre level. The process by which supply meets demand is the trip-making behaviour of consumers within the market. The approach is therefore to build a model which matches supply and demand through the allocation of flows from residential areas to financial services centres. Assuming that a region is more or less self-contained, in that people living in one region start a new relationship in that region, then the total number of new customer relationships formed at all of the outlets belonging to all of the competitors present in the region must be exactly equal to all the relationships begun by consumers across all the residential areas. In other words, the total demand within a self-contained region must ultimately equal the total revenue. This has the important implication that if a new supply outlet is opened in a region then the same number of available new relationships will be spread more thinly with a negative impact upon the existing network of outlets.

To understand the process by which customer relationships are initiated, involves understanding consumer trip-making behaviour between residential areas and outlets. To do this requires information on the residential locations of customers by outlet for all outlets in the network. Using this information, the models can be calibrated so that they accurately reproduce in detail all the possible 'flows' of customers between residential areas and service destinations or 'centres'. This shows that customer trip-making behaviour is a trade-off between the high level of provision and choice offered by large centres and the level of accessibility offered by smaller, locally based centres. Again, for some sectors, such as financial services, consumption of financial services is often subsidiary to the main purpose of consumer trip-making, which may be explained either by work or shopping patterns. This is true of both account opening and holding patterns and transactions made on these accounts.

Modelling provides a competitive edge

The particular modelling techniques used in this approach have been developed over the past twenty years in the School of Geography at the University of Leeds. Only over the past five years has this technology been exploited to its full commercial advantage. The spatial interaction model approach is powerful because it is able to handle levels of complexity exhibited by the real processes in the market which traditional modelling techniques are not designed to handle. In general, accuracy levels of plus/minus ten per cent in the revenue predictions against actual performance can now be obtained.

Having built and calibrated the spatial interaction model, it is possible to predict with considerable accuracy, the percentage of new customers 'belonging' to a particular organisation simply by knowing the locations of their outlets and those of their competitors, along with some measure of the level of service offered at each supply point. The core benefit lies in being able to quantify the impact of changes to the service delivery channels, in terms of business volumes and market share. In particular, the modelling approach can help support an analytical framework where, for example, the following questions can be asked:

- **What has been?**
  Understand the dynamics of market share by small area and outlet performance by centre over the last five years. How has this changed and why?

- **What is?**
  Understand current performance in the marketplace, nationally, regionally and locally. This would need to be supported by the development of a set of performance indicators that compare expected business levels and market share against those actually achieved.

- **What if?**
  The modelling approach allows the impact of changes in the supply or demand side to be measured. For example, the impact of new outlets can be predicted not only in terms of the business levels achieved, but also the impact on existing outlets and the catchment area can be determined. In the same way measures of the impact of competitor behaviour can also be generated.

- **What should be?**
  The 'what if?' analysis can be extended a step further by re-formulating the model so that objectives such as market share and constraints such as number of new outlets can be set and the model left to find the best solution to this problem. The model thus demonstrates the role that the control of particular supply points play within a regional market.

At a tactical level, models can be used to reduce the risks involved in investment decisions. In the DIY retailing context, for instance, growth has historically come from new store development. However, as the market moves towards saturation in many regions of the country, retailers are having to look to more and more marginal sites. In assessing new store proposals there is a break-even point below which the retailer will be unable to recover the considerable fixed costs of store development through profit contribution made by the store over an acceptable time period of usually five years. While costs are largely predictable, traditional methods of predicting store revenues have proved notoriously poor. Often the best that is achieved is a revenue forecasting accuracy of plus/minus thirty per cent compared with actual.

A typical graph of site profitability against store revenue is shown in Figure 4.24. The point (1) shows the break-even point. However, where revenue can only be predicted in advance to within thirty per cent of actual, a business case will need to predict thirty per cent over the break-even revenue level (3) if it is to be approved. In a saturated market, there are few obvious sites left that would generate this level of performance. As a consequence, retailers are finding it virtually impossible to find sufficient investment opportunities to sustain the high levels of growth that they previously enjoyed.

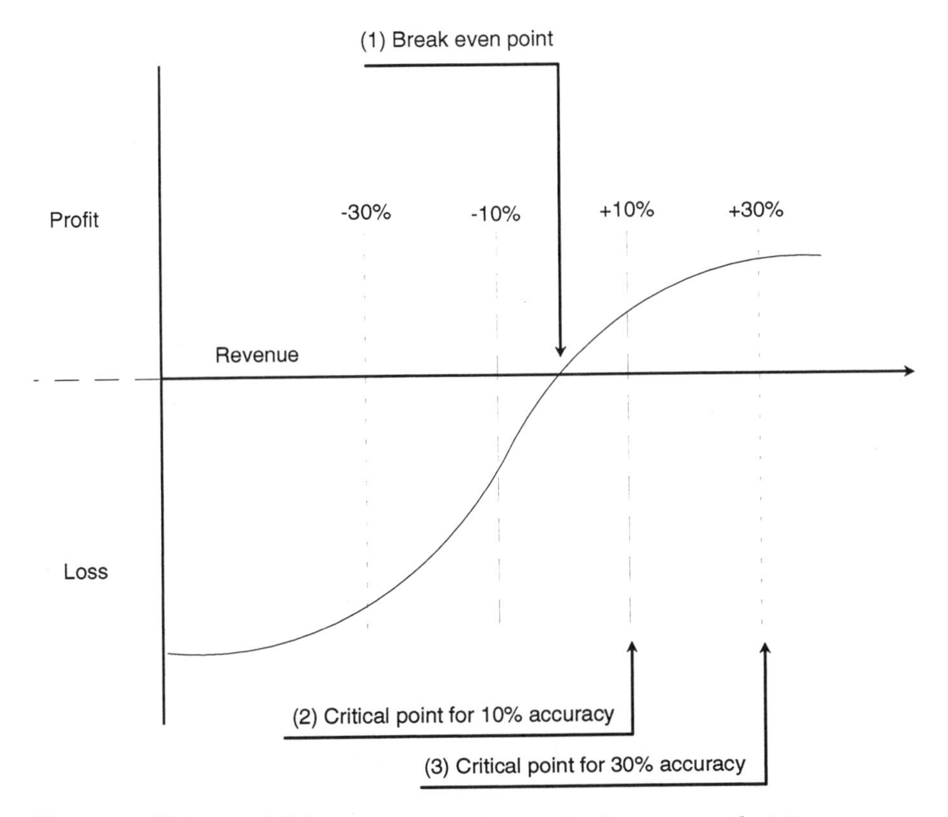

Fig. 4.24 Using model forecasts to support major investment decisions

However, by improving the accuracy of revenue forecasts to within plus/minus ten per cent, the required revenue to be forecast is reduced, the point (2). This improved accuracy allows a wider range of sites to be considered. Suppose that this enables an additional five new store openings per year over the next five years, then at an average of £5m revenue per store at a ten per cent margin this translates into additional profits of nearly £35m over the five-year period.

The interesting question is: How can companies use such an approach to re-examine and ultimately re-engineer the processes by which they initiate and support customer relationships in such a way as to produce sustainable competitive advantage?

The traditional view of marketing

In spite of the intuitive nature of the model which shows that people are increasingly less likely to visit outlets which are of greater distance from their residence or workplace, it nevertheless comes as a surprise to many on the marketing side of organisations, that rather than market share being evenly spread across a region, high levels of penetration of the market are concentrated around the outlets themselves. Conversely, there are whole tracts of residential areas where some players fail to build any significant share of the numbers of customer relationships because of gaps in their supply network, despite having competitive products. There is a widespread misconception that market share is fundamentally dependent on product pricing and other complicating factors such as brand awareness and advertising. The truth is really far more complex. Whereas, at a national level, brand strength, product competitiveness and advertising spend all play a part in determining the market share of the business, detailed analysis at the local level demonstrates that market share is most heavily determined by the location of outlets in relation to local market demand and competitor locations. This is equivalent to saying that customers determine the access points and where access is denied then no amount of marketing will overcome this barrier. As an example of this effect, even for a major clearing bank where we might consider branch-based supply to be more or less ubiquitous, given the density of their networks and their presence in almost all towns of significant size, it has been shown that market share at the local level can vary from over fifty per cent to less than five per cent between residential areas within the same region.

A company will only reap the benefits of a strong brand and competitive products where they are able to provide access to the consumer. Understanding this process is the vital key to any company's capability of re-engineering the external business process. The external part of the business process has to be the actual interaction of the consumer with the company in such a way that new customer relationships can be initiated and, once initiated, can be supported and deepened through constant interaction. The notion of customer-determined access points is thus part of the building of customer retention and advocacy.

The network effect

There exists considerable evidence to support the view that outlets do not operate in isolation. Rather, for the consumer, joining up with a particular product or service provider is analogous to taking out membership with a club. Having joined, 'members' have the right to use any of the club's facilities, hence they use services at a variety of different outlets and locations. This has clear implications for local management where they are likely to be more successful if they work as a co-ordinated team rather than in competition with each other, which is a common way of attempting to measure people at outlet level. The more outlets and wider range of distribution channels available, the greater the impact an organisation will make on a region – the whole is greater than the sum of the parts. This calls into question the traditional wisdom of managing branches as self-contained units and implies that the planning of distribution channels for local areas, self-contained areas where people live work and shop, should be co-ordinated as a local strategy.

There is a limit above which these synergies cease to apply. As the density of a network increases there will be a saturation point. Eighty per cent of the outlets in a region will not imply eighty per cent market share. This limit is the 'network threshold'. For an organisation there may be more merit in having representation in some, but not all regions of the country, but in those regions where they are present to ensure that they operate above the network threshold. As is shown in Figure 4.25, at below the network threshold, outlets tend to produce below average returns because there is not a sufficient critical mass of outlet presence to afford customers the full benefits of 'membership'. However at a certain share of the outlets, the network effect is felt and the mutual reinforcement between outlets produces synergies in the form of additional service benefits to customers. Beyond this point, and up to a ceiling, the organisation is able to achieve a higher market share than would be expected based simply on the share of supply outlets.

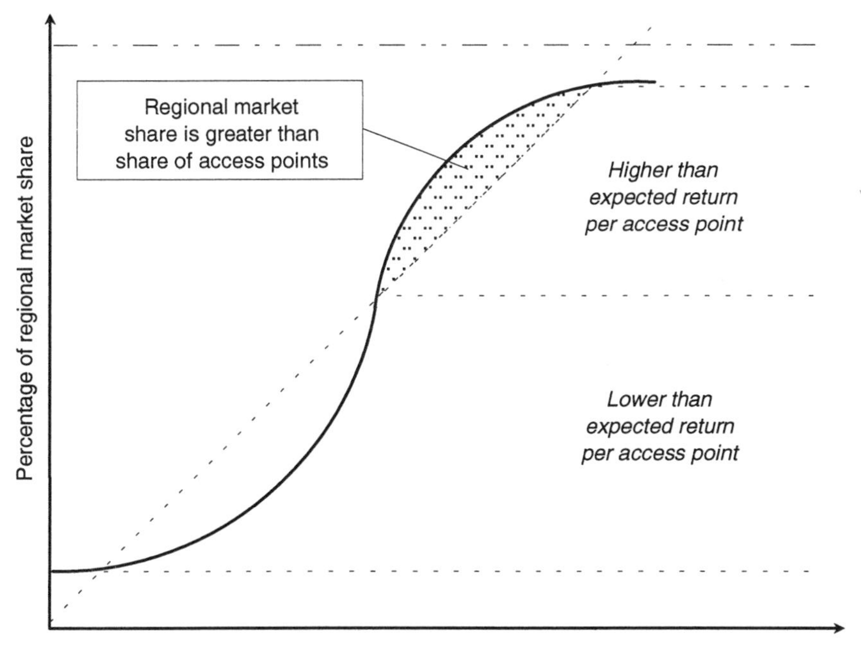

Fig. 4.25 The network effect

If the changes in societal trends are brought together with innovations in technology, and due regard is taken of the requirement to think in terms of customer-determined access points, then a whole range of possibilities can be developed to meet the emerging needs of customers. In the finance sector, for example, the types of access point that serve a local market are:

- home visit;
- conventional branch;
- other retailers;
- smart telephones;
- remote ATM;
- out of town;
- home computer;
- self-service;
- mail order;
- personal digital assistant;
- interactive multi-media;
- postal accounts.

The mechanisms of support may either come through traditional branches (but increasingly this is being supplemented) or be replaced by innovations such as remote ATMs, telephone banking and the increasing promotion of 'cashless' transactions.

Implications for companies

Companies need to re-think the way they look at the market. Rather than thinking of consumer markets nationally or even globally, the real trick is to focus on much smaller markets which correspond more closely with the actual space in which existing and potential customers live, work, shop, start relationships and make transactions. This implies analysis at the local level which in turn implies devolved, decentralised management empowered to act at the local level.

Common practice in the past has created an unhealthy, narrow focus on the performance of individual outlets. There are many examples of companies who believe that the outlet performance is mainly determined by the personality and skill level of branch managers, while ignoring the level of demand in the local market and the strength of localised competitor branches. There may well be towns in many countries where every company believes their local manager is under-performing where, in truth, all are competing in an over-supplied, saturated market where there is simply not enough business to go round. In equal measure, there are likely to be centres where the exact opposite is true.

In one company, the brand managers of products X and Y presented the targets and actual sales performance at the quarterly sales and marketing meeting. The targets were based on the national forecast demand, multiplied by its traditional national market share. This was further divided to branch level by reference to the branch sizes. The results were presented at area level to the area managers at the meeting. When product X figures were shown, every area was above target and each area manager bathed in the congratulations from the product manager. However, when the figures for product Y were revealed, most areas were below target and each area manager looked to see where they and their peers featured. Someone would get more blame than others for the poor performance.

When the branch performance was plotted against other variables the result was a straight line relationship to distance from London irrespective of branch size. In other words, the basis of the target calculation was false but the sales results were believed to be a function of branch manager and area manager ability. The quarterly meetings consistently moved the responsibility for good or bad performance to those people who could not influence sales and thus hid from view the real need to understand how demand varied and was influenced by the local market around each branch, as well as the influence on total demand of the features within each product created by the product managers. As an area manager's pay was influenced by area performance, then the whole payment methodology became a lottery with a bias created by geography. Challenging the basis of pay was perceived as a complaint from under-performers so the area managers remained victims of the system. However, if an empowered local management really are to have a positive influence on improving the performance of the business at the local level, then they need to be in possession of accurate management information of localised performance patterns within the region under their jurisdiction, and need to have appropriate decision support tools which help them manipulate the local market to the benefit of the organisation and the organisation's customers.

A re-engineered local management

In order to move away from this traditional concept of branch manager-based performance it is necessary to challenge traditional ways of managing the business. Typical local managers have not been in charge of a market at all. In fact they have been in charge of a branch. The potential 'customer' is whoever walks in through the door of the branch. The responsibilities of the manager include primarily security, administration of the office and providing good customer service and advice (to those who happen to be in the branch). However, in no way is the local manager responsible for marketing, however much organisations would like to think that this is the case.

Re-engineering branches to make them more like other retail outlets and moving much of the back-office activity off-site, does not change this. The branch remains the focus of local management and the local patterns of market share remain unchanged. To alter this requires a step change in management thinking – to make local managers responsible for their local market.

Changing the focus of local management

First, it is necessary to challenge the traditional way of thinking, which is too focused on the branch outlet. Instead the local manager must think in terms of the community or set of communities spread across the region which, when aggregated, make up the local market. Whereas branches are the traditional way of providing financial services (products, advice, transactions on existing products), branches are by no means the only way of achieving this goal.

Second, rather than the organisation being concerned about just the performance of its branches, it should widen its focus and concentrate on performance in each local market. A company should be equally as concerned about why it is getting a poor share of a particular region as it is with making sure that all branches are performing at a financially viable level. Furthermore, having established that performance varies widely within a region, a re-focused local management must be empowered to bring about desirable changes to this situation.

Modelling can help this process in at least two ways. The model can indicate what the level of company performance should be in each individual residential area within the region, and identify those residential areas where performance is below what should be expected. These areas can then be prioritised for marketing initiatives by local management. The model can also be used to test out different scenarios. Given that an organisation may be dissatisfied with its performance or market share within a region, modelling can be applied to test the effect of different configurations of access points.

The re-engineered external process

It is no longer sufficient to think of the organisation as simply the head office and the outlet locations. Rather, the new framework requires that the company moves towards thinking of the entire process which includes a link into the customer's living patterns – the locations where customers live, work and go about their day-to-day activities. The actual process by which the customer and potential customer 'join' one organisation rather than another is at least partially explained through this.

Local marketing

In order for the company to re-engineer the external process therefore, several ingredients are necessary. First, a great deal of marketing needs to be devolved to the local level. This, in effect, requires defining local areas that can be managed by one or a group of managers. Typically this may consist of an area in which five to ten traditional outlets are present. The organisation must think of this as a local market containing customers, potential customers and competitors. The local market in which the organisation must operate is fixed whereas the configuration of the organisation's access points is not. The customers' ability to 'join' the organisation can thus be viewed through the mechanism of customer-determined access points. Make it easy to join and thus lock out competitors.

Second, the local management need to be empowered. This does not mean give managers more responsibility. Rather, they literally should be given the power to make a difference. This includes not only the ability to control marketing effort through, for instance, local advertising, but, more importantly, being given the necessary management information to do the job. Local management information systems are needed and ultimately a local decision support tool which models the local market.

Making the difference

The increasing sophistication and range of possibilities available to deliver products and services to customers at a local level means that managers need to become more flexible in the way services are delivered. Simple investment economics suggests that it is not a realistic possibility to put full retail type outlets into every population centre. Instead, the local management need to think in terms of a variety of different possible access points, those points determined by customers, based on all the influences which act on customers. In this way local management is truly empowered to make a difference to the performance of the organisation and the service which it can offer its customers. Time and again it is the development of a strong network of access points and links between the consumer and the organisation which creates sustainable competitive advantage within consumer markets and ultimately determines the pattern of market share. To exploit this situation, companies need to understand the geography of their access points and empower local management with this information and that concerning customer trends, and then allow control of locally available supply.

5

CREATING A DIFFERENTIATING CUSTOMER PROPOSITION – THE VISION

'In our country,' said Alice, still panting. 'You'd generally get to somewhere else – if you ran very fast for a long time as we've been doing.'

'A slow sort of country!' said the Queen. 'Now, here, you see, it takes all the running you can do, to keep in the same place. If you want to get to somewhere else you must run at least twice as fast as that!'

Lewis Carroll

WHAT IS A DIFFERENTIATING CUSTOMER PROPOSITION (DCP)?

In BPM, the word 'vision' has to mean something tangible, the result of understanding positioning and working this through to changes in capability. In this context the vision is the desired outcome from making the BPM journey. The vision has to relate to both staff and management and requires a degree of deployment and involvement that goes beyond publishing a strategy document or an annual operating plan.

To include all these aspects, the vision is better described as the Differentiating Customer Proposition, the DCP. In this context, the DCP creates a better world for all the stakeholders; a grouping that includes customers, employees and shareholders.

The choice of words is fundamental to the changes that are required. The term Differentiating Customer Proposition is pragmatic and provides a clarity to the overall objective of everyone's job within the company. The choice of words is important:

Differentiating: Not running with the herd, but being noticed by customers and potential customers. Not being different for the sake of it, but providing a total service that puts the business ahead of the competition.

Customer: Changing the emphasis away from the business managing its prime financial ratios through forcing staff to sell, often against the best interests of customers. Customer First started the focus towards customers, but every aspect of the business's operations must be focused on serving the customer in a way that promotes customer loyalty, the route to long-term increases in profitability.

Proposition: The DCP needs to be something the customer would propose to others (advocacy), rather than the business exhorting staff to sell and forcing customers to buy. When a company gets it right for the customer then the financials follow.

Achieving a DCP has to start with a requirement to think non-functionally when looking at the business, and a requirement to think holistically when looking at the customer. The mark of success will be to make the DCP a reality in a way that nobody else does by creating a win-win partnership with customers. In this way value is added to the business by adding value to the customers' lives or business. Competitors will be left attempting to squeeze the last ounce of profitability out of their customers using the old strategies. The old strategies may have served companies well in the past but the positioning work is likely to prove that the DCP of the future will have far reaching consequences on the capabilities within the business.

Creating a DCP is the route to corporate transformation.

The three key ingredients of the DCP

It would be foolhardy to position the organisation to meet, or even create, a market need it could only fulfil with an unprofitable level of resources. Also, an enhanced capability to provide a cost-effective service that no customer wants is to have allocated resources with no hope of return. The final position and final capability deliver the DCP and the journey is the management activity that addresses both axes concurrently. However, keeping ahead of or overtaking competitors will not go unnoticed. There is an ongoing need to review constantly the future position of the business and the implications on capability. The management activity that addresses both axes then becomes the number one key business process within the organisation. This role cannot be delegated and should not remain the preserve of any functional head.

To be meaningful in a business context, the DCP has to relate both to customers and the internal organisation, in that:

The DCP needs to be a proposition to customers that excites them, retains them and has them advocating your products and services to others.

The DCP needs to include the notion of accessibility; if the proposition is that good then make it accessible to customers, convenient to them.

The DCP needs to encompass the internal framework of the entire business and all its processes; a lowest unit cost business that is also a real pleasure to work within. Above all else, the totality needs to create differentiation. The

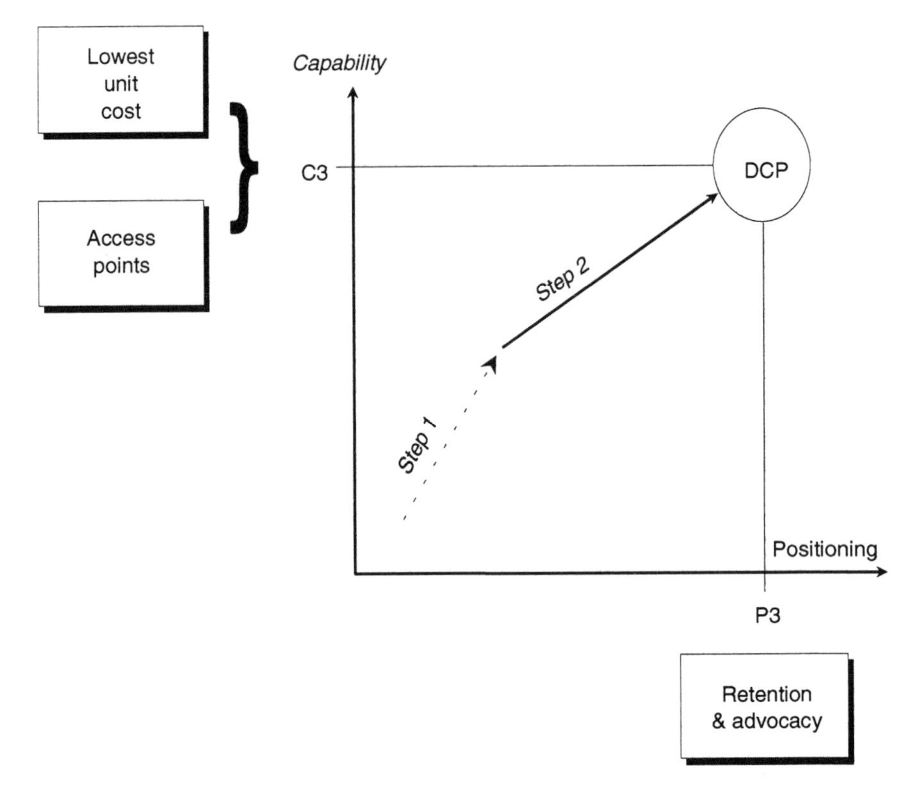

Fig. 5.1 The three ingredients of the DCP

relationship to positioning and capability is shown in Figure 5.1.

The first ingredient

The DCP must lead to customer retention and customer advocacy, providing a total service where being a customer is to value the relationship. The relationship creates a delight factor that has a value to customers greater than the sum of the traditional constituent parts of the products and services. Building a partnership relationship is thus the challenge and will require a different way of treating the customer. The retention of the customer and their advocacy more than outweighs a poor sale in the current financial year that eventually leads to permanently lost customers.

The measure of success will be:

1. *Customers saying: 'I really get value for money'; 'I trust their advice'; 'I've been with them all my life'; 'I've moved all our business to them'; 'I tell everyone they are the best'.*

2. *The cost of acquisition, replacement and servicing drastically reducing and the value of retention rapidly increasing.*

The second ingredient

The DCP must include the dimension of accessibility; access determined by customers, convenient to them, locking out competitors. Traditionally, many companies talk of distribution or delivery channels. On closer inspection, these often turn out to be mechanisms put in place for the convenience of the business. In recognising this, companies will need to look at all the places where customers are and what they may need at any time. This knowledge will determine when and how the company will always be accessible to its customers. Meeting this challenge will be another building block in growing the business.

The measure of success will be:

1. *A level of buying convenience that recognises that the buying process should not be an impediment in consumers' or companies' already busy lives.*
2. *A degree of access that poses a significant barrier to competitor entry or emulation.*

The third ingredient

The DCP must encompass the internal framework of the entire business and all its processes; a lowest unit cost business that is also a real pleasure to work within.

Conventional ways of operating and organising businesses are an impediment to making the proposition a reality for customers. It is clear that companies will have to shed many of the previous constraints that conventional structures impose on a business, put behind them feelings of functional parochialism and start again. Structural changes will be a key difference between a company and its competitors and will account for much of the success of making the DCP a reality. Using a green-field approach will be the way to overcome some of the old prejudices. The principle of lowest unit cost is established by rebuilding the organisation, starting from the customer interface, and then only including added value activities aligned in processes with minimum hierarchy, with a management framework of help and facilitation.

The measure of success will be:

1. *Achieving the lowest unit cost and therefore becoming unassailable on competitive prices.*
2. *The ability to refine constantly your own DCP as the number one key business process – a task that is owned and shared by the executive team, a task that cannot be delegated.*

Growing the market and your share of it.

Traditionally, a company may always have expected to get a certain percentage of the market. This is the case in a mature market where the competing products are undifferentiated. The market size may vary over long timescales depending on the overall economy. There may be occasional increases in market share for one company whenever they launch a discounted price. However, without a real and permanent reduction in unit costs, a discount strategy will reduce margins and prevent any investment in growth. When everyone discounts to retain market share then a shake-out among the suppliers follows, with casualties appearing shortly afterwards.

We hear much of growing one's own market share at the competitors' expense. The real objective of the DCP is to apply innovation in a way that will grow the size of the total market against the trend. By not accepting that the market has a fixed size, companies will step out of the constraint imposed by just finding out the current customers' needs. Innovation brings products and services into the market for which there was no demand expressed by customers. Innovation delights customers by presenting them with options that delight against a very low expectation from the customer's perception of the sector. Innovation works at a holistic level, bringing together a number of variables that are linked through positioning and capability.

Thinking holistically

- A bus manufacturer surveyed its customers regularly and found that the buyers were consistent in their needs. They wanted a robust, utilitarian vehicle that had high fuel efficiency and was cheap to repair. However, the users and non-users of buses, the passengers, were never sampled. A survey of bus passengers soon brought to light their low expectation that things would ever change. They saw themselves as trapped. They had to use buses so suffered what was provided.

  The bus manufacturer applied innovation in a desire to attract new passengers. Technology provided the first part of the solution. By electronically linking the buses to the bus stops, passengers were informed at the bus stops of the current position of the buses that were due. On the buses, a continuous display indicated where the bus was and where it was going. The irregular traveller, it transpired, suffered inordinate degrees of uncertainty before getting on a bus and while travelling. This was a significant factor in keeping people away from buses.

  The second factor concerned comfort. Why treat passengers so much like cattle. Why not spoil them. For the first time ever, the manufacturer invested in a number of prototypes and lent them to a bus company. The level of comfort in the design and the degree of customer care displayed by the driver were totally unexpected. People began to leave their cars at home and travel by bus. The bus company had tapped into a rich vein of new business.

- A building society ran a customer needs survey every year and found out the normal complaints from customers: queues were too long at lunchtimes, letters were hard to comprehend, and so on. However, it was aware that, although it was attractive to first-time buyers, like its competitors, it suffered a large loss of customers when a customer moved house. Clearly, customers were just hunting around for the current discounted product in the market at the time of moving.

  Here was an opportunity to increase market share by retaining customers, but could an opportunity be found to grow the market. Many people have 'too many lives' and getting into a long chain poses a lot of hassle. If a way could be found to make the whole business of buying, selling and moving easier, then more people would consider moving house. This would be increasing the size of the market.

  The building society offered a service for people who found themselves in a chain. Often one end of a chain is someone who has inherited a house and just wants the cash from the sale, and the other end is a first-time buyer. Everyone else in the chain is moving up incrementally. By part exchanging between each householder and injecting the cash into the chain at the inheritance end, the whole chain was released. By treating the chain as one big mortgage, each member of the chain then enjoys a discount.

  The art is to have something delightful happen to the members of the building society and something just short of delightful for someone who is about to obtain their mortgage from a competitor. The key is to swing someone away from just buying their mortgage on price alone, and offer something with both short-term and long-term added value. As a result, potential competitor customers then switch their mortgage supplier on entering the chain to obtain the delight effect.

  Heartened by this experience, the building society went on to develop an annual free house 'health check'. Using its valuers to give advice to householders, it provided an insurance claim service for assisting in the claim process for any policy its customers held with competitors, it offered a rolling saving and mortgage repayment deal that smoothed out the effects of variable interest rates without the heartache that attaches to fixed mortgages when their fixed term ends. The effect was constantly to bathe customers in a warm glow of help and understanding that made the company a partner in the customers' lives and led to advocacy.

  From a business perspective, the building society stopped trying to squeeze short-term profit out of luckless customers and, instead, brought more customers into the market and attracted many more away from its competitors. Acquisition and servicing costs declined and the value of retention and advocacy went straight to the bottom line.
- Many people may judge that changing the oil in a car oneself is a good money saver relative to the charges a garage makes. One garage decided to add far more value to the event as perceived by the customer. After every service the engine is steam cleaned and lacquered and the owner receives a different

car-care item to take away with them. The details of the vehicle and its complete service and repair are logged and the owners are notified in advance of any regular servicing. Also logged are the owners' personal preferences concerning the vehicle and the preferred times for vehicle collection and return. Coupled with a delightful experience when visiting the garage, in terms of staff attitude and hospitality, customers report that they enjoy the experience of their car being serviced. Not only does this garage attract business away from competitors, but it knows from its customer files that many people were first-time users of a garage service when they came to the garage. Through lifetime retention and customer advocacy the volume of business quickly pays for what may be seen initially as additional costs.

On the suggestion of one worried and busy customer, the garage now offers a service to buy the second-hand car you specify. Rather than go to back street garages or search the newspapers, you can specify the type and price of vehicle you want and the garage finds it for you. It is checked, serviced and delivered to the same standard as the service operation on which the garage's reputation was built. The cost of the search and service is covered by not carrying any stocks of depreciating vehicles. To date, no customer has declined the vehicle that has been found for them, even though the vehicle is only first seen by the customer when it is ready for collection. The size of the customer database now allows the garage to arrange many swaps between customers, where the guarantee of quality is assured.

- In one country, a survey found that twenty per cent of people regularly travel by train, seventy per cent by car and ten per cent never travel at all. The national railway company surveyed its customers and found two major complaints: customers wanted the trains to run on time and they wanted communications on the train.

  In response to the first complaint, drivers now announce to customers when the train arrives on time. This was only ever a hygiene factor; minimum expectations were met. For many customers, being told the result of a hygiene factor alerts them to consider many other hygiene factors that have not been met. In response to the second complaint, telephones are installed but the charges are high. Despite the charges being levied by the telephone company, the customer perception links the high cost to the railway company. A potential motivating factor thus backfires.

  On a radio programme one evening, a railway employee listened to a medical programme on the subject of the people's inability to step up a given distance. This condition has many contributory factors and a series of treatments were being developed. The most difficult element to treat is the feeling of embarrassment – the shame of being different and unable to get on a bus or train as the step was now too high for them. As a result, these people started to become house-bound. As these people did not travel by train, they did not feature in the customer needs survey and so were invisible. Some of these people never travelled at all, as getting into a car was difficult. Others could travel by car if someone would take them, but a train journey would have been more convenient. The railway employee saw an opportunity.

By lowering the step, market share was picked up from car travellers and increased the size of the market through attracting some of the non-travellers. However, by using a competing mode of transport, an adapted bus with a lower step, people could be collected and brought to the station – another increase in the market size. The delight factor created advocates who influenced the decisions of the able-bodied car travellers, a percentage of whom then made more frequent rail journeys.

One could argue that the effect of this one change was trivial and not worth making. However, the lesson did not go unnoticed by both the design departments or the media. The ability to think holistically brought many new ideas forward, it became a national event within the business. The media, stunned at the changes in attitudes, added more positive messages. The travelling public, delighted at the change in emphasis, became more tolerant of the occasional failure on hygiene factors.

The ladder of loyalty

The ladder of customer loyalty is shown in Figure 5.2. When people first access a company they are prospects. After a purchase they are customers. If nothing happens to upset them, they may become supporters. Up to this point, companies are vulnerable. Much cost and activity may be wasted if a prospect is not converted to becoming a customer. As a customer and at any stage in that relationship, the way the company treats customers can drive them away into the hands of a competitor. Although customers may get the same treatment somewhere else, in the meantime, they are lost. When the relationship with customers is only to the level of supporters, customers are still tempted away through competitor pricing.

The vital next step in the relationship is to make them advocates. These are customers who tell other members of the family, their friends, or other businesses, that the company is the best. Advocates stay, which is far cheaper than trying to replace customers that have left. The final step is for both parties to see each other as partners. Both partners then win as both perceive and get value. For the supplier, the key is to add value to the customers' lives or businesses. This is something different from selling them products. This is all about understanding, even predicting, a customer's real needs in an environment of total trust. This state is very fragile and likely to be eroded if the company is still using any questionable practices.

THE PROPOSITION THE CUSTOMER WILL ADVOCATE

A colleague's aged mother closed her bank account and moved her savings to a competitor. The reason? She had difficulty opening the doors into the bank as the spring was too strong. After going up and down the length of the High

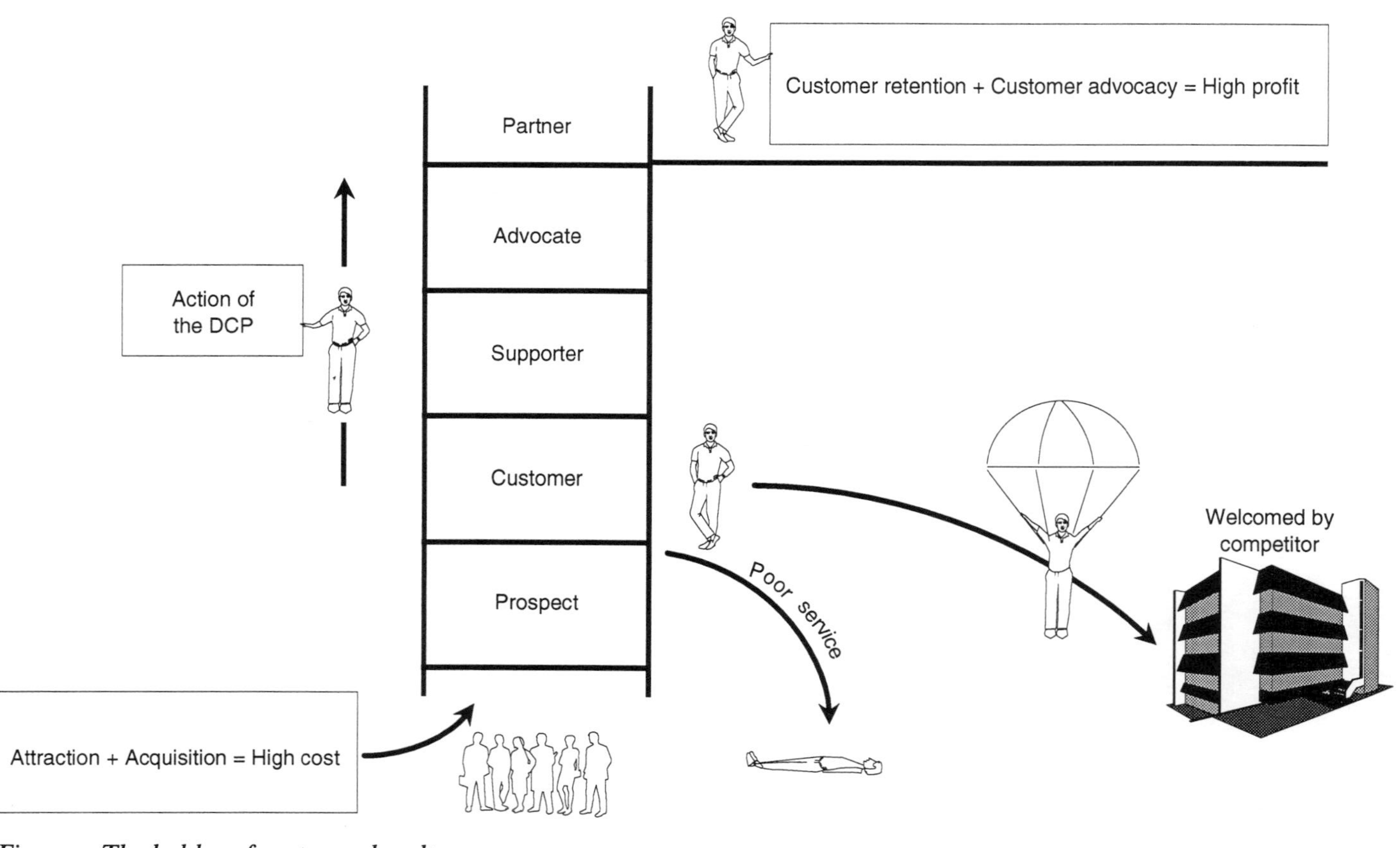

Fig. 5.2 The ladder of customer loyalty

Street, she moved her custom to the bank with the easiest doors to open. After a period of time, the latter bank noticed that it had a rush of new depositors, some with very large amounts, where the addresses were all around the same warden-assisted housing development. This was the power of advocacy. To this day, neither bank knows why they lost or gained these accounts.

A local dairy proposed to sell off its milk rounds to its milkmen. A new breed of businessperson was being created. To buy the round from the dairy, each person went to their bank manager to try to get a loan. In one branch, the bank manager saw the opportunity for the bank but had never faced this type of request from a customer before. The novelty drove his curiosity and by the end of the day a unique package was put together. At the end of the week, the bank filled up with people wearing green uniforms. It was the whole of the dairy. Nobody else had had much success overcoming the conservatism of their bank managers, all of whom had retired behind rules and guidelines from head office. By getting all of the milk rounds, the bank was able to put together an even better deal for the individuals. On hearing this, the main dairy itself switched its business to the same bank. This is the power of advocacy.

In the first example the customers chose an unusual factor by which to judge the business. In the second example, the seller had chosen to break with convention and present a proposition that was contrary to expectation. For a company to be proactive in achieving the same result it has to think beyond the confines of its individual products and services. It needs to see life through the customers' eyes.

Needs seen through the customers' eyes

A company has to ask itself a number of questions:

- What is critical that we must get right in order to be successful in the customers' eyes?
- What are our customers' 'system' needs, the way the consumer lives or the business customer operates?
- In their system of living or working, how do our products and services fit in?
- What's on their agenda with regard to us?
- What are their evaluation criteria?
- What would give our customers a better life through using us?
- Do we communicate in our language or theirs?
- Do we really make a difference or are we undifferentiated?
- Do we build relationships or just sell products and services?
- Do all of these factors change throughout the life stages of our customers?

As we saw in chapter 4, a primary key to unlocking many of the vagaries of customer behaviour is an understanding of life stages, both in the sense of a consumer or a company. To be in partnership with the customer is to accept that for richer or poorer and in sickness or in health the relationship must hold together. Both customer and supplier are locked together in one overall process. It is therefore essential that the customers' hopes, ambitions and concerns are known at each life stage. Better still, being able to predict the behaviour based on this knowledge is to be in a position to delight the customer ahead of the demand. Even better still, being known for understanding the customers better than they know themselves builds a relationship of trust and a growing dependency.

In the automotive sector, working with suppliers during the concept and design stage allows the customer (the vehicle assembler) to take advantage of the emerging technology at the suppliers, rather than attempting to source parts already designed on the basis of old knowledge. Bringing the assembly plant (the internal customer) in at the design stage (the supplier) allows the needs of efficient assembly to be brought in at the beginning of the launch cycle, rather than the assembly plant bemoaning a repetition of poor designs. It has taken many years for the automotive sector to change its ways and reap the benefits of simultaneous engineering. Breaking down functional parochialism, misplaced pride and the traditional lack of trust in all the customer/supplier relationships have been the key barriers to overcome.

Again in chapter 4 we saw that societal trends shape customer behaviour and also that one sector can change the perceptions by which customers measure products, services and relationships in another sector. A stunning knowledge of one's own products and the technical brilliance that differentiates them from those of one's competitors may be of no interest to the customer if the judgement criteria is based on an entirely different premise.

A company selling commercial vehicle tractor units made much of the technical features in both its literature and its sales training. The real needs of the customers were met when a supplier was able to help develop their logistics strategy over the next ten years, indicating how a fleet of the new tractor units and trailers would play a key role in growing both businesses. Unfortunately for the traditional supplier of vehicles, this ability was non-existent so the business slipped away. The needs seen through the customers' eyes had changed as they moved into a different life stage, prompted by their own customers, the supermarket chains, which also introduced a different set of evaluation criteria. This example also illustrates how the process chains are getting ever longer. For many companies, a knowledge of the customers' customers is now a prime requirement.

Building components to build a proposition that the customer will advocate

The holistic nature of the proposition now starts to develop. Looking out at the world from within the business, companies quite naturally refer to their 'products' or their 'services'. Over the years, catalogues are been produced showing all the products in glossy brochures. Internally, computer systems have been designed and put in place to serve a product uniquely – a different system for each product. In the finance sector this has particularly been the case and now accounts for a customer perception that the supplier's left hand does not know what its right hand is doing. In fact, with the explosion of product proliferation, it is now like receiving mailshots from an octopus with unlimited access to its own paper-making factory and a vocabulary of words no one else ever uses.

In all the companies we work with we ban the word 'products'. We prefer to use the word 'components' where a component is anything that touches the customer. The sum total of all the components builds to become the total perception the customer has of the business. Components would include all the traditional products in a technical sense, but also need to include giving information or advice, the nature of any contact for whatever reason, the fixed assets and premises that customers enter, what they read in the press, and the attitudes and behaviours of staff they meet. The implications of this simple change in nomenclature are profound.

With this change, companies are no longer in business to sell products. Rather, they will be creating a proposition which builds components together at any life stage to add value to people's or companies' lives such that they see the company as partners in their lives and becomes advocates as a result. The societal trend that demands that customers be treated as individuals, a segment of one in a customer base of millions, exacerbates the problem for traditional businesses and highlights the opportunity for those who realise they need to create a DCP.

Some examples of building components into the proposition the customer will advocate are now appearing. One car supplier now enables customers to select from a range of components those that together make up a car. Sitting at a screen with the customer, major components are selected to form the vehicle. Bodies, engines, trim, drive options, glass types, fuel type, ranges of accessories, colours, payment methods, delivery options, trade-in calculations, and so on. The customers can design their own cars from the components to produce the vehicle that is right for them.

Elements are worked on to make the components. For the vehicle, the companies' internal competences will be focused on specific expertise to work with steel, rubber, plastic and so on. The internal capability creates the cost, but also another opportunity to delight and differentiate as delivery of the new car is never more than five days later. Even the access points have been determined by the customer. For a standard desktop or portable computer, the company supplies a disk so customers can build the proposition in the comfort of their homes without a salesperson having to guide them through a difficult brochure. Letting customers use a modem allows customers to avoid what they may perceive to be an uncomfortable selling environment.

We can think of any company building components in the same way. In Figure 5.3 we are shown how in the automotive sector the components are built to create the DCP. Similarly, finance sector components can be built to create a proposition at each life-stage event. The key life-stage events all centre around changing the mix of savings, borrowings, insurance, advice, information, planning and payment transactions. The proposition the customer will see becomes 'the financial package which is right for the customer at each life stage'. The proposition is not a product in the conventional sense. It is what customers understand the company to be in the context of their lives, and leads to a change in capability in the way customers are handled that the competition will not be able to match.

	Automotive	*Finance*
Elements	Iron Chromium Carbon Hydrogen	Knowledge Money Data
	↓	↓
Components	Wheels Facia panel	Savings Lending Insurance Advice
	↓	↓
The differentiating customer proposition – 'the branding'	The vehicle which is right for you, built for you to meet YOUR needs. We know how to make travel a real pleasure.	The financial package which is right for you, built for you to meet YOUR life-stage needs. We know how to create peace of mind.

Fig. 5.3 Building components to make a DCP

The key is to think innovatively to build the various components together in a way that benefits the customer and the company and locks out the competition. Every employee should be brought into exercising their minds on creating the overall propositions which may be valid to support the DCP. At this stage, they would only be ideas resulting from a burst of innovation. However, customer-facing staff are generally very good at knowing how to build components together, once they are released from the shackles of rules and mistrust that so often characterise the environment in which they have to work. The freedom to do this (supported by appropriate systems) will be true empowerment used in the correct sense of the word.

Components are not just re-badged products. Every customer contact is a component of the proposition. With this comes the need to empower customer-facing staff so that they can use their own judgement to create delight and differentiation contrary to customer expectation. This will be the real challenge for a company – to deliver this combination to customers at any time in their lives and, further, at places determined by the customers.

Actions speak louder than words

Many companies have attempted to achieve a truly customer-facing business. Brave as these attempts have been, few have achieved the final state of high levels of customer retention and highly vocal customer advocacy. One company, Nordstrom, a chain of retail stores in America, has gone a long way to achieve this. Collecting 'Nordstrom' stories has become something of a pastime among the company's customer–advocates. So many stories exist that some may be apocryphal. Here are a few examples:

The free trainers shoes

Someone walking through a Nordstrom store was stopped by a salesperson as the shopper was wearing a pair of tacky old trainers. On being questioned, the shopper replied that they were purchased in Nordstrom's four years before. Without a further word, the shopper was taken to the sports department and was given a new pair for free.

The point of this story is not that companies should give away everything for free. However, to delight customers against their expectation is to create an advocate for life who will leverage opinions in such a way as to create long-term sales that recover the cost of one pair of trainers many times over. In Nordstroms, the level of staff empowerment to make such on-the-spot decisions is very high.

In your own company, what 'free trainers' could you give away where the cost is minimal relative to the value of retention and advocacy?

The made-to-measure suits

A man was thoroughly fed up with hearing how good Nordstrom was from his wife and daughter. He decided to fix them by ordering a suit on a Thursday for completion the following Monday. He was pleased to discover on the Monday that his suits were not quite ready and departed for his three-day conference. On returning to his hotel room the following day, he found his new suits, plus two pairs of shirts and matching ties. He now knows more Nordstrom stories than his wife and daughter.

A change from sceptic to advocate in one move. Again, local empowerment and a clear policy of doing things to gain retention and advocacy pay off in the long term.

In your own company, does a recovery action still leave a disgruntled customer or a delighted and retained advocate?

The complete outfits

Three spring outfits were delivered to a customer's home, complete with matching make-up, shoes and accessories. The customer's 'personal shopper' included a note saying that Nordstrom hoped that they had read the needs of the customer correctly in their choice of outfits, and would the customer care to try everything on to see if any one of the outfits were suitable. Should they not be suitable, then ring on the free phone and the goods would be collected. The delighted husband of the customer rang later to say they were so pleased with all the outfits they couldn't make a choice. As they would be keeping all the outfits could they be charged to their account.

This example shows how the customer relationship had progressed to one of 'partnership' with advice you can trust. The personal shopper concept also overcame the problem of finding time for shopping for the products.

In your own company, do you know your customers better than they know themselves, and is your advice then trusted and acted upon?

Does all this seem a bit farfetched and inapplicable? The pundits, sceptics and cynics in Wall Street all waited for Nordstrom to fall flat on its face. The reality is that the company has far outstripped the performance of its competitors in its sector with a much lower expenditure on advertising costs. Customer retention and advocacy is a low unit cost way of increasing profitability. For Nordstrom, their motto, 'Respond to unreasonable requests', and a refusal to have organisation charts and job descriptions, has created a very different type of company. Employees live the motto and management do everything they can to remove the conventional barriers.

Not all such examples track across to a particular business. However, the principles of creating the customer relationship do apply, as do the lessons on achieving a degree of empowerment that continually reinforces and builds a better and better DCP. Where the value of the proposition is high, the price of the conventional products and services becomes less of an issue and may even attract premium pricing. A valuable relationship is worth buying, both from the customer's and the supplier's perspective.

Lifeguiding

For consumers, a long-term relationship through customer retention means that during this time the customer will go through a series of life-stage events, such as buying the first house, raising a family, changing jobs, retiring, and so on. At each event a different requirement from a company occurs. The art is to predict these events by knowing the customer, or similar customers and, in an atmosphere of trust, meet and exceed the customer's expectations. Through this process, the depth of the relationship increases as more members of the family, and a more complex set of life-stage events, all come to the company to have their needs met.

From a customer's perspective, individuals go through life faced with a number of hopes, ambitions and concerns that determine how customers will form a judgement on the action they take at any time. For some sectors, such as finance, the company has to compete with a customer perception that buying toothpaste is seen as more interesting than obtaining life insurance. Such companies need to recognise that in the busy lives of customers, the time available to deal with the supplier is a key issue in determining personal priorities.

Clearly the role of customer facing staff will be changing, as will any form of customer contact from the company. To recognise this issue, we coined the description 'Lifeguiding'. Sometimes this will be an individual, the person with whom the customer is dealing in a sales environment, but the reality for the business is that everyone and every system is lifeguiding if they are contacting customers. From the customers' perspective, the company should be seen as the customers' 'Lifeguide'. Although the principle is easier to visualise in a consumer-related business, it works equally well in a business-to-business environment.

The implications of getting the right competences to be able to lifeguide will have a profound effect on how a company views its own people. In a sense, everyone will have to have an ever increasing set of competences to be ever better lifeguides. A lifeguide becomes the top job in the company, and everyone should be seen as contributing to lifeguiding.

To achieve lifeguiding means that the staff at any customer interface need to see their primary objective as achieving this result. To allow this to happen means removing all the shackles from staff that currently prevent this relationship building up. Also, technology, internal processes and management support to staff must be geared to achieve the same result. In Business Process Management these factors are the primary drivers to build capability.

MEASURES INFLUENCE BEHAVIOUR TOWARDS (OR AWAY FROM) THE DCP

What gets measured gets done. Old measures will have been successful at delivering the old strategies. If the new strategy is to deliver the DCP then new measures are required to deliver the new strategy. Although this is recognised by most companies that want to make the transition, there always appears to be a reluctance to drop many of the old measures that have served the company well in the past. The wrong measures, or the right measures misused, are the strongest influence on employee behaviour. If we assume that nobody comes to work to be disruptive or destructive on purpose then we can presume the following

1. Managers are themselves measured in a way that leads them to believe that they are controlling their staff's behaviour in a way that is in the best interests of the company.
2. Staff react to those measures that are linked to appraisals, money, advancement and the avoidance of 'poor performance' against the measure. The measures on staff are often in conflict with meeting the needs of customers, let alone building a DCP, but without empowerment and in an environment where management is not readily questioned, staff will meet the needs of their immediate customer (their manager) and ignore the needs of the external customer.

If managers never realise the long-term damage an inappropriate measure is doing to the business, then can we blame them for driving wrong behaviour? It is interesting to note that often the measures placed on staff are a combination of the staff member's output, competence and performance, whereas the manager is measured purely on the degree to which output targets are achieved by staff. The competence and performance of a manager should include a knowledge of how measures determine behaviour. If this was the case, many staff (victims) would be freed from the conflicts they are subjected to at the customer interfaces.

The balance between old and new world thinking is shown in Figure 5.4. The two are balanced around management attitude, which is in turn influenced by their awareness of how measures influence behaviour. We can also imagine the tiny figure of a customer-facing staff member running to either side of the fulcrum, driven in either direction by signals from the company. Staff members would be drawn to the new world after participating in a 'customer-first' workshop. They would then be seen scurrying back to the old world after having been asked by their manager if they had met their cross-selling targets at the end of the month.

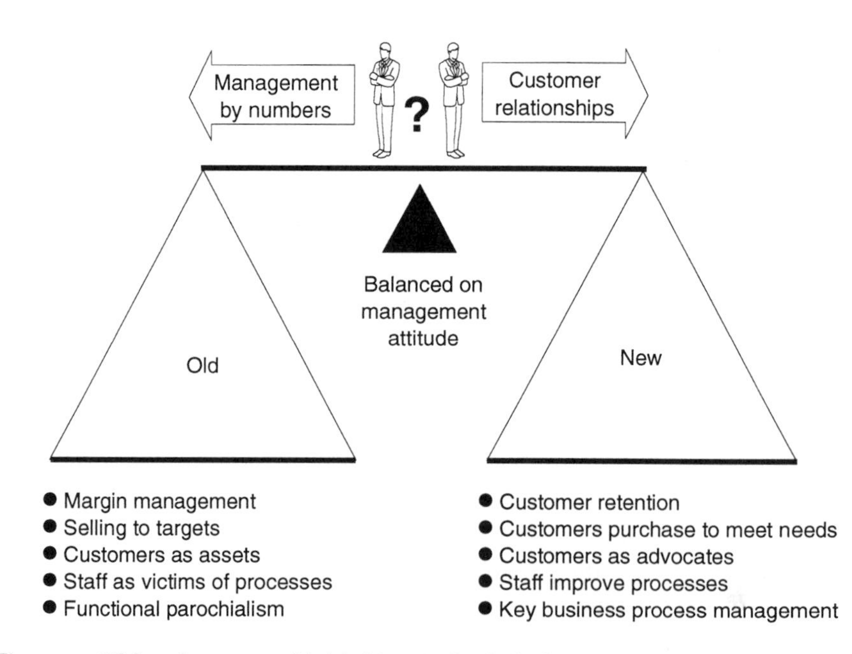

Fig. 5.4 Old and new world thinking is finely balanced

The old world and new world measures are mutually exclusive

The power of inappropriate measures to prevent the delivery of the DCP can be found in many companies. Branch selling and retail chain operations demonstrate this issue clearly. The example is from a fictional finance company.

To obtain the sales numbers required in the short term and to deliver the right cost–income ratios and other measures, often meant that staff compromised their relationship with customers. At branch level, few actions could influence sales. The key influences were external to branch control. Typically, these were:

- Macro-economics had created a decline in the overall market, even though the company had maintained its share of this market.
- Research showed that there was a communication lag of two years before the company's branding had a significant effect on consumer reaction.
- Regional demographics had the effect of creating different levels of sales at branches that were notionally identical. Sales in one region always outstripped sales in other regions, irrespective of area or branch efforts.
- Competitor activity in terms of the number and quality of the outlets that were in the proximity of the company's branches were a clear determinant of business levels.

The external influences on branch performance were the key factors to be considered. Despite this, the figure for the total national sales opportunity, calculated from a basis of the current strategy plus the key ratios needed to satisfy the City, were split arbitrarily across all the areas and branches in a way that did not recognise the local conditions. The arbitrary number, a target, was used later to measure branch performance.

Achieving the target (even where this was below the true potential derived through considering the DCP), led to the following branch perceptions:

- a belief that the branch had some influence on the level of sales;
- a re-inforcement of this belief by issuing prizes (vouchers, bottles of wine, trips to Paris);
- getting to the top of a ranked list in competition with one's peer branches (rather than the real competition) was the key to bonus and promotions;
- belief that the target system was good for building one's career (though the target system was contrary to building long-term customer relationships).

Not achieving the target led to the following branch perceptions:

- belief that the branch has failed to achieve;
- a re-inforcement of this belief by poor appraisals;
- being at the bottom of a ranked list prevented promotion;
- everyone thinking that it was a mathematical possibility to get 'above average' on the ranked list.

In this type of environment, sharing best practice became a meeting at which branch managers attempted to share an arbitrary target less arbitrarily, rather than the group meeting to understand what would help to attract customers (advocacy of other customers) and keep them (by providing a differentiating service for their lifetimes). A half hour with any branch team brought out the intellectual expertise that was used to 'fix' any target the branch had been set, although managers refused to believe that the target system was driving this type of behaviour. None of these activities added value to the customer and were performed at great expense to the company. This ingenuity should have been directed at delivering and evolving the DCP at every place where there was customer contact.

As each branch was unique on the basis that sales were determined by the population it served, ranking the performance of one branch against another was meaningless. No branch was significantly better than any other due to its own efforts. By eliminating internal competition, everyone was freed to help each other for the benefit of the company as a whole. In fact, helping each other became recognised as a key measure, and the ability to provide help became a key competence.

The key is to achieve a WIN-WIN-WIN relationship between the customer, the staff and the company. This is achieved by delivering the DCP, retaining customers, increasing market share, and growing the market.

Changing the basis of measurement

In the old world, a typical sales scenario would be the following:

Customer:	'I'm interested in X' – the enquiry.
Salesperson:	'What you really need is Y – the sale.
Salesperson:	'and it's a condition that you also have Z' – the cross-sell.
Customer (later):	'Why have I got all these expensive products?' – the decision not to go there again.
Salesperson:	'I've met my target for the month' – promotion.

To achieve customer retention and customer advocacy, the two mechanisms to deliver increased profitability, means that the new world scenario should be the following:

Lifeguide:	'I think we agree you will need X now and I will check with you again in six months to see if Y is more appropriate. It is better if you get Z from this other company we recommend.'
Advocate:	'I get value for money and always trust their advice. I have been with them all my life and I tell all my friends they are the best.'

If conventional measures are generally contrary to the DCP, then the new measures will need to be something to do with knowing that the company is achieving the DCP. Typically, these will take the form of the following questions:

- Are we growing the size of the market?
- Are we growing market share by taking custom from our competitors?

- How many new customers in the sector are now with the company?
- How many of the customer relationships can be considered as partners?
- Are customers taking more components?
- Are we retaining our customers for longer, and for all of their lives?
- Do customers trust us?
- Are customers telling other people that the company is the 'best'?

Clearly, these are more difficult measures and it will be some time before companies see an upward trend. However, if the DCP is the strategy to grow the business and increase profitability, the primary measures will be those that indicate that they are putting the DCP in place. This will be to do with the speed that the capability is changing and the degree to which the company has transformed.

Companies making these changes often start with a concern that the new measures are loose and esoteric. We would contend that the new measures allow the business to be more in control of its destiny. On closer inspection, many of the measures conventionally used in companies have been put in place, or have been subsequently misused, to exercise control over people rather than to promote meeting the needs of the customers. A strong case therefore exists to make a point of stopping old measures rather than attempting to run with the old and new at the same time.

ACCESS POINTS

Earlier in the chapter it was suggested that 'selling products' should be replaced by 'building components' as the route to achieve differentiation, retention and advocacy. Similarly, we would suggest that the expression 'distribution or delivery channel' be dropped and replaced by 'customer determined access points', or 'CDAPs' for short. This will force the company to view its own business through the eyes of the customer and place the proposition the company is offering in the context of the customer's own hopes and concerns, and the time the customer has available to deal with the company. For this reason, CDAPs are part of creating the overall DCP.

The changing customer perception

After bus deregulation, cities and towns filled up with new types of buses. They were smaller and ran more frequently. They had curious names such as 'City Hoppers'. On many of the routes a new type of differentiating service was introduced. Quite simply, the bus would stop wherever a passenger wanted to be set down or picked up along the route. In one instant, we had a perfect example of customer determined access points (CDAPs) and customers were freed from having to go to a bus stop, an access point determined by someone else.

In Hong Kong, the service has gone one step further. The road down from Victoria Peak, the high ground residential area that looks down over the city, was becoming congested with vehicles. To put on more buses was to add to the congestion and buses ran to a timetable that meant waiting for a bus to be sure of catching it. The answer was, an eight hundred metre system of twenty escalators and three travolators. Commuters travel down between six and ten in the morning and the system reverses between ten in the morning and ten in the evening. Customers can now determine their access points both in the distance down the hill that they enter the system and the time they want to use the service. Unlike buses travolators are not locked into a schedule, so waiting is eliminated. It is expected that the service will attract 26,000 people per day.

Both these examples indicate the power of CDAPs to provide a better service and entice customers.

Confusing the customer

Access points are a critical consideration for the consumer. Most of my bank transactions (withdrawing cash) are done at my local supermarket. They have a large convenient car park, the shop is pleasant, and the opening hours match the time I have available. At the checkout one has only to pay for a few items and then ask for an amount of cash. Handing over the debit card completes the transaction without the need to remember a personal identification number. Of course, the temptation to buy other non-essential items is always strong so the supermarket makes an additional gain.

On the other hand, even after completing three customer needs surveys, my bank has not moved its branch to a more convenient location. On my rare visits to the bank I have not been tempted to obtain any non-essential financial products and no fruit or vegetables were on display. We can also imagine the frustration customers have when finding branches closed or undermanned. How many times have they taken the trouble to park miles away and trudge through the high street to join a long queue of people just wanting to do a cash transaction or get a savings book amended? The last thing customers want to do is then hold up the queue to discuss their financial concerns.

In the finance sector, in particular, history has left companies with quite separate product streams which have been broken out into separate divisions with separate objectives, targets and bottom-line responsibilities. The customer can be at the end of a number of distribution or delivery channels and gets somewhat confused as a result. In the case of a typical building society, this could mean: receiving advice from the financial planning arm that is allowed by law to select from any competing product provider on the basis of best advice; receiving advice from its estate agency chain that can supply mortgages from competitors; or from its own branches that can give best advice only from the selection of its own products; or from its mobile investment advisers that advise across anyone's product; withdrawing cash from its ATMs; taking out insurance policies from head office or unsecured lending from another head office department, issuing credit cards from another division; and receiving the direct mail showering down from all of them. Is it any surprise that the customer is somewhat confused in the middle? Not only is one company competing with itself for more of the customer's 'wallet-share', but each distribution channel has its own access points determined by the business.

New opportunities to delight and differentiate

Where companies are particularly vulnerable is to have their customers influenced by another sector so that the customer receives new evaluation criteria. In the past, garages would fight many price wars to attract business. With the emergence of out-of-town supermarkets, new shopping patterns have emerged where convenient access by car to the point of sale satisfies a customer need. From this it was a small step to add a fuel-filling service. Such outlets account for a large proportion of the fuel sold. The cheaper fuel prices attract more people to the out-of-town location and into the supermarket.

In response to this challenge, conventional garages have attempted to transform themselves into mini supermarkets. However, where out-of-town shopping locations separate their customers into two streams, conventional garages mix their customers at the same pay points – a source of delays for vehicles waiting to use vacated filling points. It is all too easy for potential customers to drive on to another garage. The astute user of the supermarket can also practice parallel processing. One member of the family going round the shop while the other fills up the car with fuel.

From the company's perspective, the design of distribution and delivery channels is generally guided by the cost of the activities to distribute the products or provide a service. When viewed holistically, the element of customer determination in the design might give a solution that has a higher cost relative to the former calculation. However, if adding in the notion of CDAPs leads to customer retention and advocacy then the long-term impact on revenue may well outweigh the additional cost. A simple example exists with the emergence of pizza home deliveries.

Advances in technology also continue to bring the companies closer to their customers. For some businesses, technology can bring the company virtually uniquely in front of each customer. In chapter 4 we saw the influence that societal trends will have on customer behaviour, particularly the cocooning effect and the 1990s child. Computers are common in most companies and now appear in many people's homes, and will be an increasing access point for customers to interact with suppliers. The launch of TV channels devoted entirely to selling products and services is a CDAP for a growing segment of the population.

Similarly, interactive multi-media (IMM) offers customers the choice of searching at their own pace, without interference, for information that meets their particular needs. The products and services on offer no longer have to be sited at the conventional outlets, but can instead be sited at locations convenient to customers where they will take greater pleasure in accessing whatever is on offer. As a by-product, IMM software tracks the customers' interrogation sequences and can thus be further fine tuned to match the customers' needs.

A retail outlet chain for home furniture and appliances believed that it understood what customers wanted to know and so arranged its stores and the manner in which staff provided help in a predetermined way. The feedback from the users of the IMMs brought to light a quite different set of needs. The store layouts have since been changed and staff are trained to have a quite different interaction with the customers.

Where the positioning work highlights that differentiation, retention and advocacy will come from changing from pressure selling to giving advice in an atmosphere of trust, this may have a significant effect on the business to be conducted at the customer interface. For a finance company, where customers may need to discuss their hopes and concerns, then businesses will need to create environments where this can happen. This may not be possible in, say, a conventional branch in a high street. The location and environment would need to change, as would the type of interaction of staff with customers.

For your own business, a useful exercise is to brainstorm every possible way of providing access to your products and services and then all the ways customers may want to have access. Having done this, repeat the exercise but this time consider every possible customer interaction with the business and treat these and the products and services as components that are building the proposition that customers will advocate. In the latter case, thinking holistically will start to unearth a new set of possibilities. This exercise, completed by a finance company, is shown in Figure 5.5.

Locking out competitors.

Careful consideration of CDAPs can help to lock out competitors. The point is not to create an unsurmountable wall to prevent customers from leaving you as they will resent this type of action when the gloss of the proposition wears off. Rather, it should be something that has value to the customer but still retains the choice for the customer to go elsewhere. The point is thus to ensure that you remain the preferred choice through a differentiating CDAP.

A manufacturer of buses had a spares manufacture and supply operation which was based on the customers having to keep necessary spares in their own stores. Urgent items were manufactured in small batches. The costs were high, as were the prices. Wherever possible, the company would mark-up the price of proprietary items to recover costs on manufactured parts. This policy was based on expecting customers always to accept that spares are expensive. Eventually, customers took the trouble to order proprietary items direct from the manufacturers or other third-party suppliers. In the latter category, some suppliers began to offer a manufacturing service for manufactured parts.

To meet this threat, the bus manufacturer developed a system to log the complete build and modification history of every vehicle. The customer

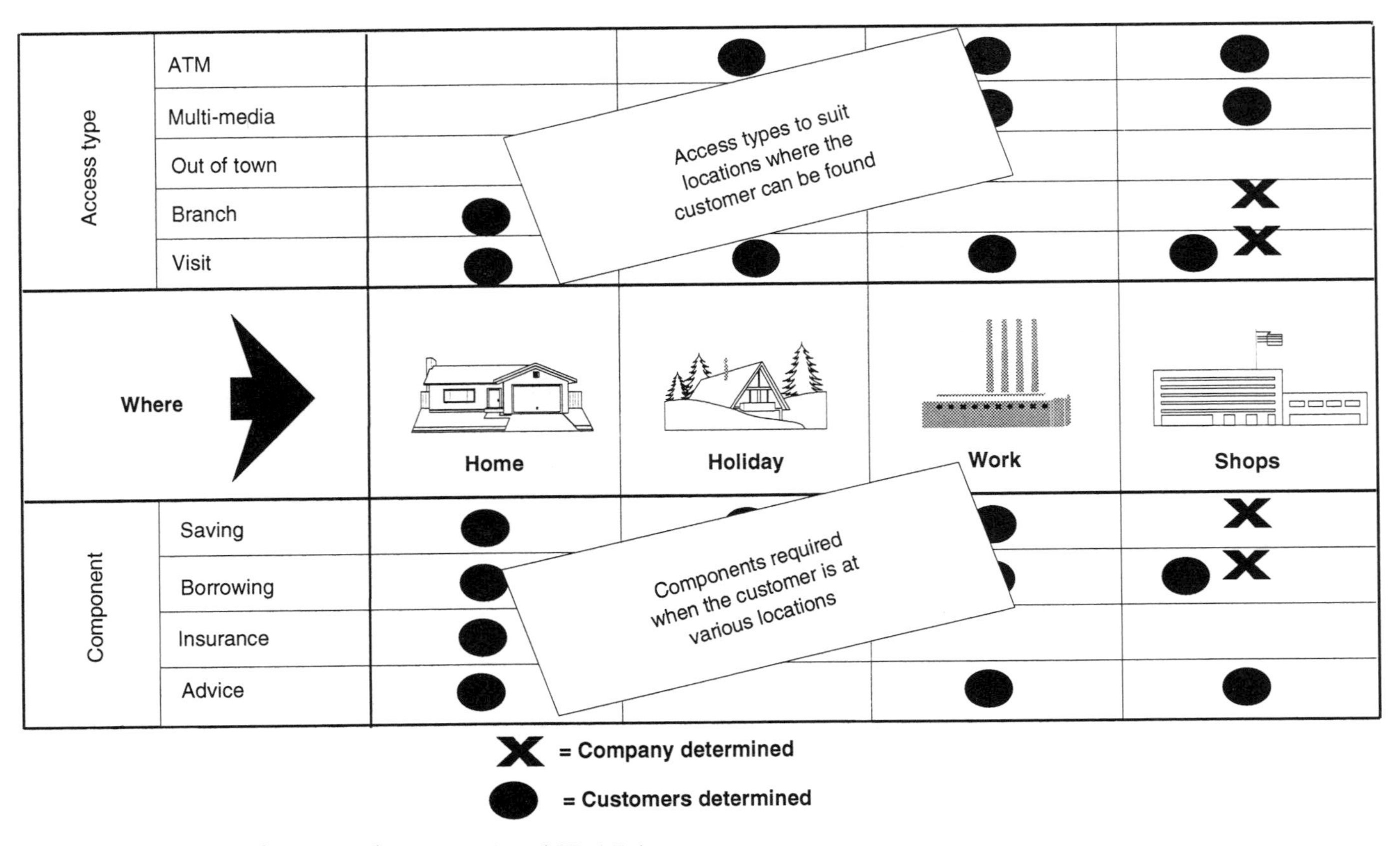

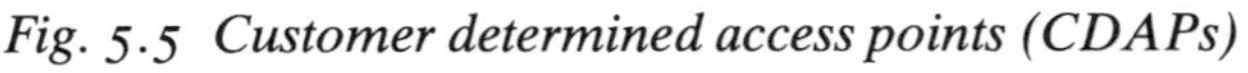

Fig. 5.5 Customer determined access points (CDAPs)

entered a unique vehicle identifier into the system at their own site and flagged the parts required (or area damaged if caused by an accident). They then ordered spares via the terminal. The bus manufacturer stepped up the service levels on parts supply so that the customers could reduce their stockholding to virtually nil and thus free up space for better primary uses in their central garages. The supplier updated the delivery progress as an aid to scheduling the repair work at the customer. A further enhancement was to provide a mobile accident and repair advice service. A skilled assessor would go to the customers' garages, assess the work, interrogate the system and order the parts.

Traditionally, the spares part of the business was seen by customers as a rip-off and by the manufacturer as a separate division with bottom-line targets to meet. The change meant that a holistic view bathed the customers in a warm glow where they valued the total service. The change brought retention, an important factor when whole fleets of buses are being replaced. Any competitors were locked out if they could only meet the initial needs of supplying just the vehicles.

A building society wanted to provide country-wide access to its current and potential customers. It also realised that in the process of buying a house, the purchaser is likely to go first to an estate agent. Also, estate agents could act as independent suppliers of mortgages and other insurance related products. The building society saw that this was one of the reasons for losing its customers when they made their first house move; customers moved to an area where they were not represented or the estate agent moved the finance business to another provider.

As part of its development of a proposition that customers would advocate, it also considered the issue of CDAPs. It knew that setting up new conventional branches was expensive and not in line with meeting the emerging customer needs. It also knew that to change the customers' habits of a lifetime by cutting out the estate agent part of the process was unlikely to be successful. The answer was to think holistically to find a solution that made everyone a winner – the building society, the estate agent and most of all, the customer.

The company found a number of small independent 'local' estate agents which would provide the increased number of national access points. It selected and further trained the estate agents who had the same values and beliefs in relation to customer retention and advocacy. As the building society's reputation grew so did the value of the 'kite-mark' on the independent agent. For the agent, not always perceived as being customer oriented, the branding was itself a mark of quality.

The agent was supplied with an interactive multi-media database system on which details of all 'kite-marked' agents' properties are held throughout the country, along with 'geographic modelling' information, additional locality information and annual Home Health Check information on the property. The geographic modelling system enables a buyer to select a house in another town by postcode. A preferred postcode location in their own town selects properties of the postcode type in any other nominated town or area. The first stages of the selection and buying process can be performed without travelling to the other location. The customer and linked agents both benefit from the advanced interactive database.

In an earlier section of this chapter we saw at Nordstrom how the 'free training shoes' story illustrated the value the company put on achieving a lifetime relationship. The building society chose to pay the customers' costs for selling the house, even if, at that point, the customer was with another society. The agent thus concentrated on its core activity, selling houses. In return, the agent allocated space for the society to provide its 'lifeguide' service for advice for all the life-stage events of anyone that came into the premises. The lifeguides enjoyed the full IT support from the centre for all customer-related information.

Modern ATMs provided cash transaction facilities. In the past, conventional branches were designed mainly for this purpose. As transaction methods became more sophisticated, the customers' need for lifeguiding services became the priority and this required a different type of environment. Agent-based lifeguides were also mobile, providing their service at people's homes, where this was the preferred CDAP. The use of lap-top computers completed the link between customers and the main databases.

By taking a long view, driven by the need to achieve customer retention and advocacy, the company developed an overall holistic set of services so that the sum of all the components, including access points, grew in value from the customer's perspective. Suppliers of other services, such as estate agents, were seen as partners in the whole chain of delivery of the DCP. As a result, competitors became locked out of the total supply chain as their were no weak points to gain entry. By avoiding the previous inclination to see customers as trapped and therefore vulnerable to be cross-sold to, the company chose to put no obstacles in the way of a customer choosing to exit from the relationship. This in itself was a differentiator.

A LOWEST UNIT COST BUSINESS

The third ingredient of the DCP is the achievement of a lowest unit cost business. During the transformation of the business to achieve the future position, the point P3 on the BPM journey, the company is vulnerable on three fronts:

1. It needs to have processes that will deliver the proposition the customers advocate through access points the customers have determined.
2. The branding of the company, and its products and services raises customer expectations which should not then conflict with experiencing the company through the attitudes, beliefs and behaviours of the staff at the customer interfaces.
3. It is vital to avoid an inappropriate short-term reaction to a competitor initiative by always being price competitive.

We hear much of the need for 'management commitment' to a programme of transformation. We hear much of the need to flatten structures and increase spans of control. We hear much of the need to instill leadership into managers and reduce management by control and fear. Many initiatives wither on the vine because they approach the issue as isolated attempts to change management behaviour and organisational methods that have existed for generations.

The work on positioning, if done thoroughly, will inevitably bring the business to the conclusion that the old ways were valid in the past, and did well for the company, but the old ways are invalid in the context of the external changes in the future. Prior to moving on to Step 2 of the BPM process, the company will have to set down the fundamental conclusions it has learnt from the positioning work, then list the implications this will have on capability, the point C3 on the BPM journey map.

The key driver for changing the capability from C2 to C3 is the achievement of lowest unit cost. People can argue the for and against of cultural change at a human level, or the for and against of function versus process. No one can argue that lowest unit cost is not good for the business.

Here the distinction between cost cutting and lowest unit cost must be drawn. Anyone can do cost cutting. It is arbitrary, short-term, damaging and requires no management skill whatsoever. Achieving lowest unit is subtle, requires nerve, is radical and innovatory, throws out tradition, requires science and great management skill. Cost cutting by itself leads to business decline. Achieving lowest unit cost leads to growth in the market and your share of it.

Capability issues arising from the creation of the DCP

What has become apparent is that the conventional way of operating and organising businesses is an impediment to making the proposition a reality for the customers. It is clear that companies will have to shed many of the previous notions of organising the business, put behind feelings of functional parochialism and start again.

Unbundling structures

The delivery of products and services to customers is the end result of co-ordinated activities by different groups working within the business. Organisations are hierarchical, and quite subtly, the structure determines who really is the customer. It is the person you work for, the person who directs your activities, the person who appraises your performance. Hierarchical structures, and the measures that go with them, have taught staff to do as they are told by those above them, even where this is in conflict with meeting the needs of the source of revenue. The traditional management structure causes managers to put functional needs above those of the multi-functional processes to which their departments contribute. This results in departments competing for resources and blaming one another for the company's inexplicable and continuing failure to meet and exceed customers' needs efficiently.

Unlearning the past

Management will need to unlearn many of the practices they have used for decades so that they and their staff can be freed from the cultural baggage that will prevent corporate transformation. People learn how to get on by perpetuating the role models set by their managers and directors; the old role models are no longer appropriate. Competing with one's peers to gain attention by the old rules puts staff in conflict with the need to generate higher overall business performance. All barriers to working together in teams must be dismantled. People can and should be trusted and not suffer continuous audit of their decisions. Using irrelevant data to measure someone's performance, just because something is simple to measure, must be replaced by measures that indicate whether the DCP is being achieved. People should be asked to do things within their competence and not personally suffer the consequences of failure when tasked beyond their competence. They should challenge the long and accepted norms of the structure and way of running the business, however heretical, and this will not be a career-blocking factor.

Uncoupling career paths

As traditional structures are pyramidal, then by definition not everyone can get promotion – a recipe to frustrate half the people that seek advancement. Where 'career' means striving for status through position with a reducing opportunity to get there, the goal is satisfied for many by moving to another company.

In future, everyone should be able to get promotion while keeping a flat structure by building on their key competences to better understand and service customers. Career then means having the hygiene factor of pay met through increasing competences, and motivation for everyone coming from having pride in workmanship and recognition within a team. The need is to have teams of competences working together, responding flexibly to the emerging needs of the business, as it makes its never-ending journey on positioning and capability.

Uninhibited employees

Procedures put in place in the interests of standardisation and efficiency can confuse some customers and leave others feeling patronised. The implications of positioning will have a profound effect on the nature and degree of empowerment that staff have to deliver the proposition where the specific delivery of the positioning will have to respect the various customer segments. The competences of staff may be out of line with the future need to create a different relationship with customers.

'Customer Care' or 'Putting people first' was aimed at changing the behaviour of people in regular customer contact. Customer expectations were raised, but the processes failed to deliver consistently. Staff know the issues but lack the empowerment to influence the processes outside their control that would truly create care for the customer. Staff will need help and new techniques, they will need different degrees of empowerment, and they will need the support of the entire organisation through its structures and business processes.

Unprovoked, now led

Companies will need to develop another way of addressing conventional hierarchies. Those people we might recognise as 'managers' will have a different role. In the future, such people will have the role of leaders. This is not another call for exhortation or simply training. The beliefs and values, and consequent behaviours of managers, and a change in their role from control to one of helping others, has been found to be the factors that can leverage the change inside a business. A new role for managers is a key change in the delivery of a change in capability to deliver positioning.

A leader is someone you can think of as having those competences that can look at the processes within a group of staff to ensure that the mix of competences in the group is always matched to the current delivery of the DCP, and to the future capability implications of the evolving future positioning. Leaders interact with each other to share best practice, they provide help and facilitation to the people doing the work. In a sense, the people doing the work could appraise the leader on the basis of the leader's ability to add value to the processes in which the people work. Ideally, we would expect that the team would always recognise if the leader lacked the competences of leadership and the leader would also be helped to obtain the necessary skills.

This turns on its head the notion that a manager is there to check on whether the people have done the work. Conventional appraisals should become a thing of the past, as should performance-related pay. Lack of capability is rarely the issue. Having insufficient competences to do the tasks often is. People cannot perform what they are expected to do if they lack the competences. Management provocation and motivation through individual performance-related pay assumes that people are inherently lazy and work-shy. Leadership means an end to fear-driven behaviour and the start of applying motivation in a way that people desire to learn and work together in the best interests of the customer and thus of the company.

Beating people with 'sticks' used fear as the weapon. Offering 'carrots' used money as the motivator. Beating people with carrots used performance-related pay as the means to confuse and set staff against each other. After corporate transformation, the willing will be led by the able. Any other combination will not achieve lowest unit cost.

THE FUTURE BUSINESS DEFINITION AND DEPLOYMENT

So far in this chapter, the work on positioning, P3, and the implications on changing capability, C3, will be pointing towards a degree of corporate transformation that the business may not have experienced before. Even though all the staff may have already been involved in Step 1 of the journey, issuing an edict to make fundamental change is unlikely to obtain immediate buy-in. Up to this point, staff and managers may not have been able or had cause to question the direction of the business and the need for more radical change. Step 1 would also have raised a degree of expectation in the eyes of the employees that their views have value and that they have some degree of empowerment. In recognition of this, the executive team should readily see that to involve managers and staff in developing the change in capability to C3 is to add value to the implementation. For people to get involved means informing them of the broad changes that the positioning conclusions have determined. This process is called *policy deployment.*

Policy deployment

Policy deployment in companies will become a fundamental way of getting change speedily into the business in a manner that ensures that everyone has contributed and is 'singing from the same song sheet'. There is nothing clever about policy deployment. The people in the top team will have applied their specialist competences and those of staff to create the positioning, P3. However, in some businesses, the observed behaviour has been for each functional head to research their own functional topics concerning positioning and then unilaterally to put in place initiatives to effect change. Without top-team consensus, this runs the risk that some initiatives will be mutually exclusive. Advertising that the business is differentiated from its competitors but failing to change the capability of the business will lead to a customer experience that is below expectation. Creating a demand that the customer has difficulty in getting access to wastes the resources devoted to attracting the customers in the first place. Having a high-cost capability leaves the business vulnerable to competitor attacks on prices. No one should have the authority unilaterally to put in place 'functional' initiatives.

Deploying the initial plan, P3/C3

The overall stages are shown in Figure 5.6. The conclusions made by the top team will be the preliminary plans. These plans could be called the 'definition of the future enterprise' and need to contain a summary of the positioning work and its conclusions, a summary of the implications on capability, and a clear signal concerning the values, beliefs and behaviours that the future enterprise will be working towards. The latter issues will have been brought into sharper focus if the views of staff and managers had been specifically measured during Step 1 using an attitude survey or similar technique. The plans should then be deployed throughout the business in a series of interactive workshops rather than sending people a copy of the document. The ground rules for deployment are:

1. The documents must be written and any deployment delivered in a language all staff can understand. Where elements of fear exist or staff are not used to asking questions, then their silence could easily and wrongly be taken as their complete comprehension of the messages being deployed.
2. The whole deployment process must emphasise explaining things to people and not prescribing the solution. Prescription is likely to be the old way of doing things. Changing the format of deployment provides a clear behavioural and cultural signal of change from the top team.
3. On principle, the process should have the guideline that staff and managers have a mandatory right to question the validity of the explanation being given to them.

In some poor cultural environments, these first signs of openness and a willingness to open up a dialogue with staff can be risky. Staff may use the process to test the executive team's resolve and true commitment. Chief executives should not be surprised to find themselves facing a department that has high levels of sickness and a sudden increase in volume of work, having to field questions where the staff want to know when any new recruits are joining. If, at that time, there is a corporate block on recruitment, there is no way that any individual case would be noticed at executive level or be treated as special in anyway. The fine words in the deployment exercise will sound bitter against the realities that some staff are facing on a daily basis. Despite this sort of testing and probing, in general, the majority of staff will report that they welcomed the opportunity to get involved in the early stages of implementation and that they see how their role in development of the detail is vital for the plan's success.

Boulders on the runway of implementation

In a company characterised by functional heads presenting their annual operating plans in isolation from the consequences of the plans on others, coupled with a high degree of cross-functional competitiveness and 'points scoring', one is likely to see behaviours in the top team that are contrary to delivering the DCP. It is not unusual to find managers who are keeping 'boulders' behind their back in order to hurl them on the runway of a colleague's implementation plan with disastrous consequences. In policy deployment we are trying to achieve the opposite effect; the removal of boulders on the runway of implementation.

At each level of the organisation and from every functional and expert perspective, people are required to absorb, understand and evaluate the consequences of the plan on their own area of knowledge. The high-level positioning

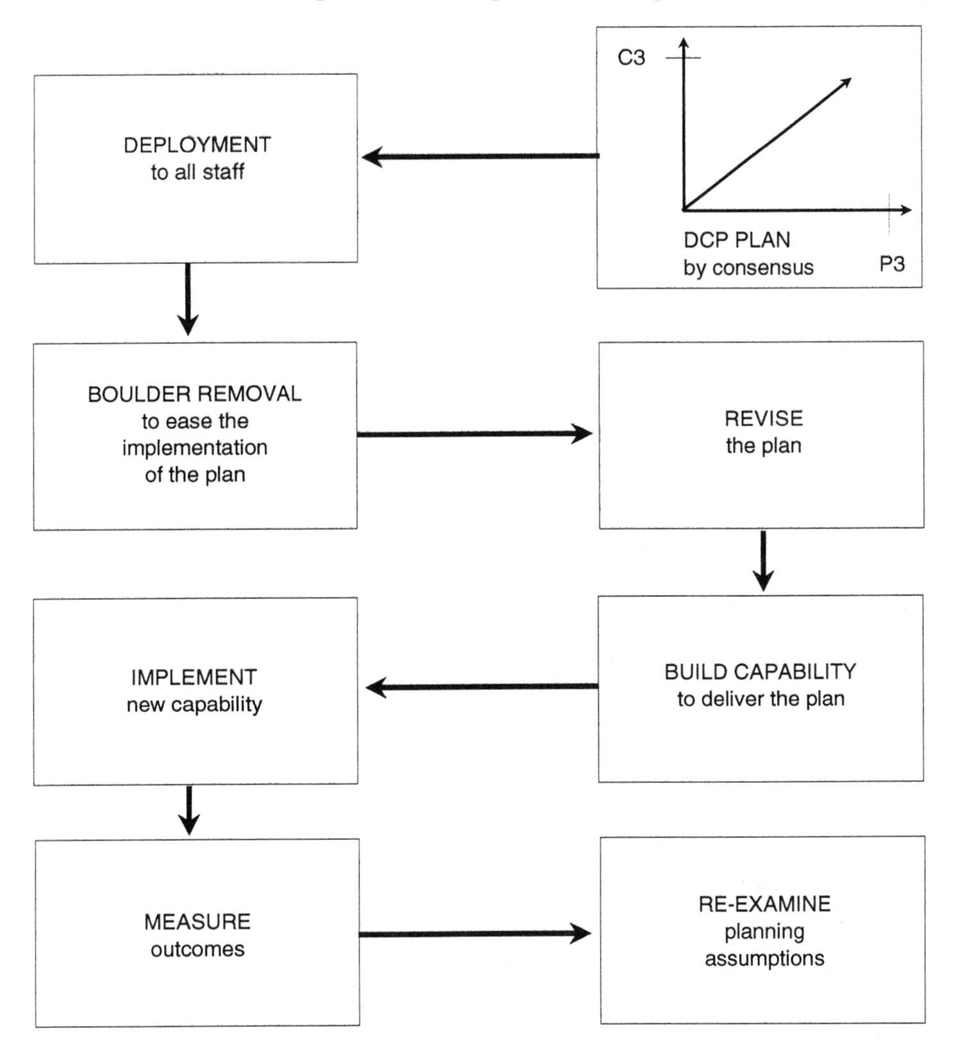

Fig. 5.6 Policy deployment

work and the implications on change to C3 will not have considered every nuance of the plan's implementation. During this iteration of deployment, the need is to find features in the plan that could not be delivered unless a specific action or variable is considered and addressed quickly. A typical boulder could be something like a mismatch of competences that pointed to a training need, or a requirement for some a specific type of technological development. Without this iteration, the boulders would appear later when implementation actions had put many changes on the critical path. A boulder at this point then stops everything.

At the local levels within the business, and particularly near to the customer facing activities, staff will be able to add more of the current positioning issues, P2, and how these are linking to the P3 issues highlighted in the plan. Without this bridge, staff will find it hard to reconcile the changes for C2 and the issues raised in moving to C3. The deployment process becomes a fast iteration that leads to a revised plan that can be implemented. As everyone will have contributed and many boulders will be removed, companies will be able to react to any emerging threat or opportunity in the marketplace.

Building the capability, C3

Having agreed the revised plan, subsequent to the deployment and feedback process, the task is to build the necessary capability to achieve the whole of the differentiating customer proposition. The capability changes will be to do with creating CDAPS and achieving a lowest unit cost internal framework. A model around which to develop C3 is outlined in chapter 6. However, the key issue to consider is the control of the capability building teams and, in particular, the need to ensure that concurrent change is keeping in step. Any new process that falls out of step creates a discontinuity somewhere in the business and puts the overall plan at risk. Keeping the build capability programmes in step is too serious an issue for it to be delegated to a department. So often one can find corporate project offices being put in the position of trying to tie together the disparate functional initiatives. Other functions see the corporate project office as just interference, and the project office is reduced to just plotting other people's progress on a spreadsheet that is circulated once a month.

Outcomes at variance to the plan

It will be no surprise that the plan will have financials in it. After all, positioning creates the volumes and revenues, while capability creates the costs. Costs and revenue are consequences of the plan. In future, by measuring the outcome of the plan, any outcome that is different from that which was expected should lead the company to question the assumptions and conclusions that created the plan and the degree to which new capability was put in place effectively. This is the continuously improving planning and implementation process.

The implications of this approach are that companies can stop linking failure to achieve an outcome as being a failure on the part of the people doing the work. If a process is predetermined and the people come to work to do a good job with the competences they have, then failures are generally not due to individual performance. In the past, staff have been more likely to be castigated for failure to deliver a poor plan rather than a director being castigated for creating a poor plan. In the future, all measures need to be used to focus on improvement rather than as a weapon against employees at any level. Even the process of corporate transformation can be improved through continuous learning. This directly matches the values, beliefs and behaviours that the company will have deployed throughout the business – values and beliefs the company can honestly share with its customers.

Breaking free from the annual planning and budgeting cycle

While building the changes in capability to achieve C3, the implementation will be following a stable path. Two events break this stability. In the first case, this would be associated with competitor action, either directly through announcing a discounted price for a product or service, or through a sudden change in the customers' perceptions and evaluation criteria (e.g. the launch of a special limited edition motor vehicle or the launch of a new money-back or vehicle exchange guarantee). In the second case, this would be associated with the build capability teams getting out of synchronisation from the agreed plan, creating a need to re-balance resources or timing (e.g. developing the launch of an advertising campaign while the IT hardware is suffering a delay in implementation at the customer interface).

Both cases prompt a need for the top team to address the arising issues and to assess the implications. We call this ringing the 'P3 alarm bell' or 'blowing the C3 whistle'. On hearing either of these from any source at any time, the top team convene to consider the next steps. This response is not only quick, but it breaks the traditional cycles that drive most company behaviour, such as peering at monthly accounts and sales figures, quarterly strategy reviews, annual operating plans and budgets. Any changes to the current stable state are then communicated using the policy deployment process to look again for boulders on the runway and to involve staff in the reasons for changes to current plans. The overall process is shown in Figure 5.7. In this way, the organisation is able to optimise the use of its resources between looking out on positioning issues, building new capability, and delivering the DCP on a daily basis. It will also enjoy having the most adaptive and responsive mechanism to meet any emerging threat and still achieve lowest unit cost.

A DIFFERENT COMPANY

All this seems like a very different type of company from many that exist today. Can a company achieve such a change? A business exists because of its customers and its own people. Survival is not compulsory, and everyone will have to work hard at getting the changes right. That is, as right as it can be at the moment.

Companies need to become learning organisations. Further, external forces dictate that companies must change and constantly strive to keep ahead or overtake the competition. As the company changes, the competition will notice and, in turn, will respond to the threat. The way that businesses will describe their own DCP will lead to radical and innovatory changes. Even after implementation, companies will constantly need to apply continuous improvement to the DCP itself in order to always be the 'best', whatever the customers determine the 'best' to be.

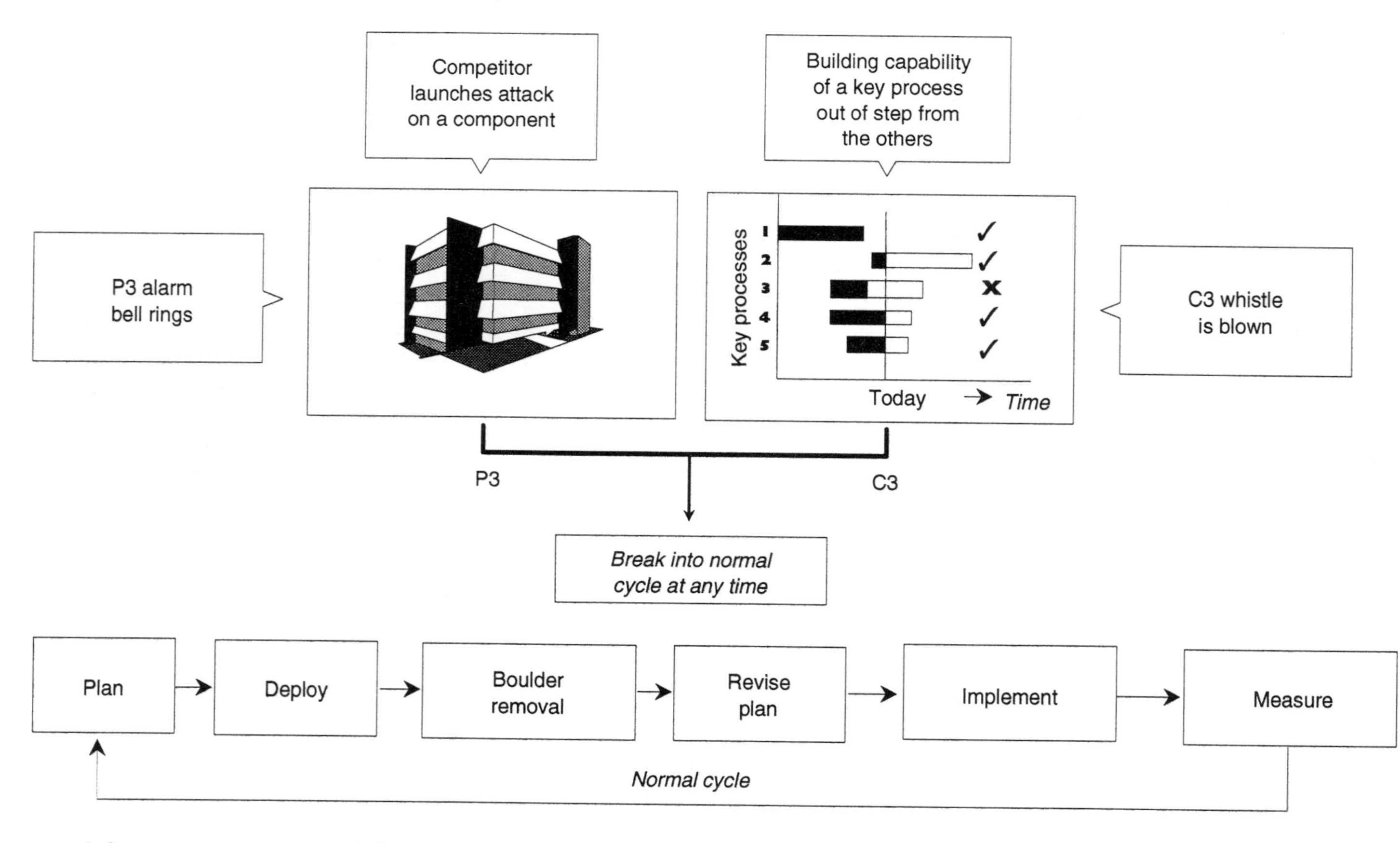

Fig. 5·7 A fast response process of change

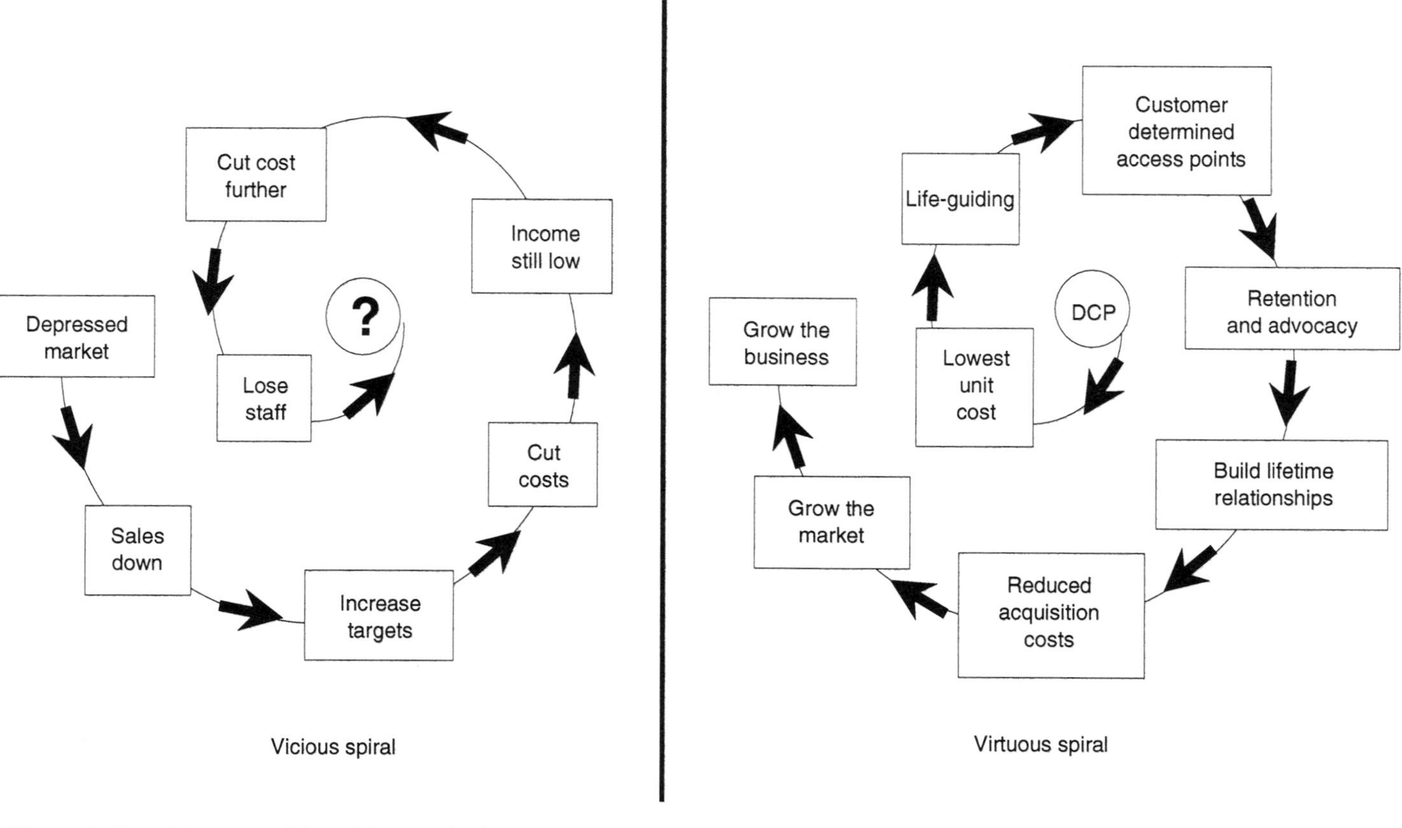

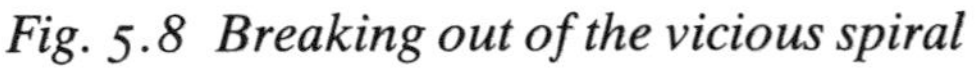
Fig. 5.8 Breaking out of the vicious spiral

In the future, the company's results will provide the proof, and the people in the business will confirm, that the company is managing to get it right.

Changing the spirals

Depressed market conditions coupled with old strategies cause a vicious spiral of cost cutting plus pressure on achieving sales through distribution and delivery channels. Although for many companies, high profits are still being enjoyed, some key measures (retention and advocacy) could well be moving in the wrong direction. At some point in time the spiral of cost cutting will bite into a business's ability to generate revenue through the conventional approaches. Mergers and acquisitions may only spell a brief reprieve and a once-off adjustment to cost–income ratios. As is shown in Figure 5.8, the DCP breaks out of the vicious spiral and creates a positive spiral of growth.

6

THE BPM JOURNEY: Step 2 – Radical and innovatory changes to capability

In the new climate, doing nothing was itself an act of opposition. Because by doing nothing we change nothing. And by changing nothing we hang on to what we understand, even if it is the bars of our own gaol.

John le Carré

DESIGNING A LOWEST UNIT COST INTERNAL FRAMEWORK

Why lowest unit cost? When this is achieved, the business will also be price competitive. That is not to say that the business has to reduce prices because it has reduced its unit costs. On the contrary, the proposition the customers advocate may have a value that is greater than the original products and services that were offered. On this basis, a premium price may be possible while still growing market share and growing the size of the market. Achieving lowest unit cost allows the business to react to competitor threats through pricing. However, where a competitor may reduce prices to buy market share for a short while, this action will not be sustainable in the long term. The business can react quickly to such pricing threats without sustaining damage in the long term. This could also be used as a weapon against competitors, knowing that a reduction in your prices can be sustained well beyond the point where a competitor would start to suffer permanent damage through the reduced income levels.

One can argue that the end result of creating a lowest unit cost business is to have a company that looks very much like one that has approached corporate transformation through implementing Total Quality. However, various surveys have pointed to the success rate of TQ initiatives being as low as thirty per cent of those that take the TQ route. There is evidence to suggest that where a TQ initiative is predicated on the chief executive and the executive team making a personal transformation to become true believers in terms of their attitudes, beliefs and behaviours, any lack of true conversion remains visible. The organisation then treats the whole programme of change with the usual degree of scepticism and cynicism.

Using the positioning and capability model and the creation of the DCP as the vehicle for corporate transformation means that nobody will argue with the notion that a lowest unit cost company is good for growing the business. However, to achieve this state is to drive towards a different way of organising the business that will challenge nearly all conventions and traditional principles. But if a company is truly set on making the DCP a reality, then it should work through the logic, at minimum as an exercise in free-thinking, and then test this against its own resolve to make it to the end of the BPM journey. If the company balks at any stage, then it will know it retains some residual vulnerability to competitor attack. Rushing headlong at changing everything at once in a short timescale brings its own risks. Keeping one's resolve and working at a pace that ensures corporate transformation at a digestible rate is the best guarantee of success.

The basic principles

Building a lowest unit cost internal organisation framework starts bottom-up in terms of structure and inwards from the customer. Such a framework creates the greatest challenge to any conventionally structured business. The lowest unit cost business has little hierarchy, delivers the proposition that customers will advocate, through access points the customers determine, with staff working in robust processes helped by leaders facilitating the application of the tools and techniques of continuous improvement. Staff will have the competences to do the tasks and will be given the competences to meet the evolving needs of the business in relation to its customers. The previous central functions and head office roles will become self-adapting and concerned with balancing their resources between watching out for future changes (evolving positioning) and implementing process changes (evolving capability). The business will be using policy deployment as the mechanism to involve everyone in constantly delivering the evolving DCP, and the business will have a rapid response to competitor threats, rapidly changing customer perceptions, or unsynchronised development of capability. Measures will drive the DCP and its improvement and will not be used to make victims of employees or create behaviours contrary to the needs of customers.

Competences are the foundations of the new structure

Many organisations, or more particularly the human resources and personnel functions, are moving towards using competences on which to base job descriptions, staff evaluations, pay and performance rating – in other words, the spark of transformation snuffed out by misuse on invalid measures. It can also be argued that the source of this movement, one function, risks the change being seen as just another functional and parochial initiative. More paperwork and more confusion just to satisfy the personnel department.

However, where a change to using competences is still inside an untransformed business or one that is not even contemplating the journey, then competences becomes just another misused measure and does little to get to the heart of the issue. The move to use competences should only be considered as part of the delivery of the DCP and the requirement to achieve lowest unit cost. In any other context, introducing competences will just add another layer of bureaucracy and complexity. It will become an industry all of itself.

Toppling the hierarchy

In all companies embarking on this stage of the BPM journey, their own organisational structures will be the end result of many years of working in a particular way, with managers who have built their careers within the 'rules' the company has laid down. This poses a significant barrier even to getting managers to consider a different basis on which to build the structure, and much time can be wasted while managers worry whether they will retain status, position and pay, as measured within the current rules. The answer then is to start with different rules.

A classical structure is shown in Figure 6.1. In such a structure, career means striving for status with a reducing opportunity to get there and therefore a high incidence of people moving to other companies to get the position that provides money and status.

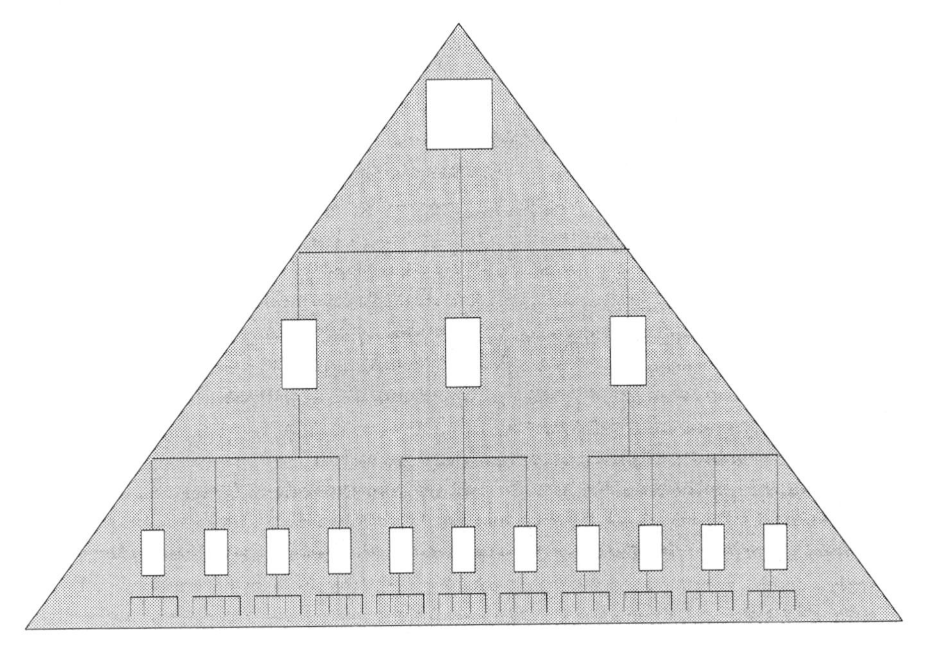

Fig. 6.1 A classical structure

In the public sector, organisation charts are known as the 'establishment'. Vacancies are treated as meeting the establishment and many instances of union disputes and negotiations have centred around whether a vacant position on the chart can be removed. The very nature of the word establishment seems to set everything in stone, oblivious to the changing needs of internal and external customers. In one nationalised railway organisation, the absence of a manager for a period triggered a procedure whereby a subordinate took the manager's place and received a temporary increment of higher grade duty pay. The hole that was left was then filled by yet another subordinate. The organisational hole cascaded down the business and across the country until eventually a line maintenance person became a temporary supervisor and the points froze one morning on a busy intersection of three key commuter lines.

Referring again to Figure 6.1, we could treat the area of the organisation triangle as the total value added to the business. As you move up the chart, each layer is paid more and the numbers of managers in each layer declines. On a broad scale, each layer's cost can be the same, with each layer, in theory, adding equal value. However, if the current task of many managers is to deal with process failures, firefighting and fixing mistakes, then we can question whether each layer is adding equal value, particularly if the existence of the layers and the inter-functional barriers are themselves causing the failures that managers find themselves working on to resolve. The structure starts to feed on itself and it exists because it exists. This can be exacerbated by some of the measure and reward systems that are still used where a rise in salary could only be awarded if someone managed others and manager's pay was determined by the variety of the tasks managed. This proved to be a recipe to create diversionary activities and tasks determined by the manager, and an increase in the number of layers, in some cases up to twenty-seven.

The starting point

The starting point for a lowest unit cost business is shown in Figure 6.2. Here we start with a mix of people doing a set of tasks using the best methods with a mix of appropriate competences. The hygiene factor of pay is met through the competences they have, which will be visible to all. Their motivation comes from being able to take pride in what they do and recognition from their peers within the teams they are working in. Everyone comes to work to do a good job within their total capability. The 'performance' of the people is not an issue and therefore appraisals on this basis are not required. There will be no need to rank people one against another, so there will not exist the notion of people being above or below average. The whole purpose of the team is to maximise the sum of their competences and use measures to improve the processes in which they work and to measure that their outputs are creating the proposition the customers will advocate.

If people are allowed to grow their competences in the direction their inclination takes them, then they remain motivated and at peak performance.

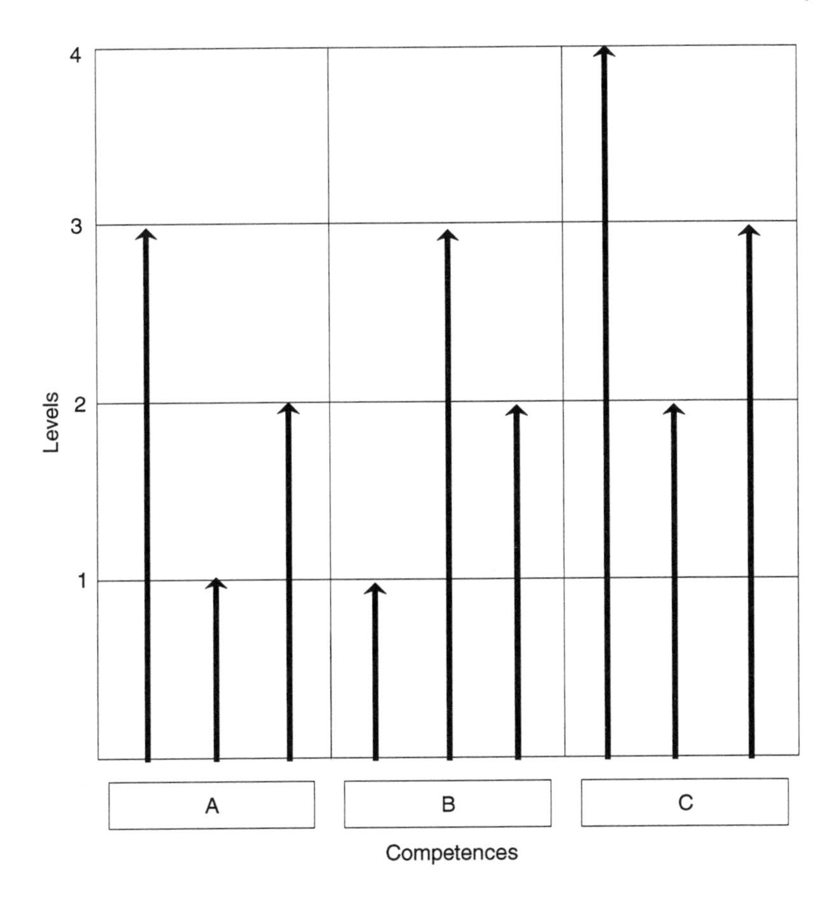

Fig. 6.2 The starting point for a new structure

In most cases, the growth in competences will be aligned to the needs of the business, after all, they are doing tasks they enjoy in the first place. Overall, the growth in competences represents growth and expansion of the business, each person being able to add more value which in turn reflects in their pay. The value they add is in enabling the business to reduce both the cost of doing any task and in being better able to delight and differentiate the company from its competitors. As cost and revenue are being addressed at the same time, then the value added makes the increased reward affordable.

At the limit, it could be argued that everyone will grow to have competences well beyond the needs of the business. This situation would not arise overnight and everyone will have a personal limit they would set on competency growth. However, a high value-adding business is one that is growing its market share and growing the size of the market. At some point it may well question why it imposes an artificial constraint of remaining in its own niche market or sector. If the delivery of the DCP is well understood within the business then it will have an overall capability to delight and differentiate at lowest unit cost in whatever field it chooses to operate in. The business can then spawn new companies in perpetuity; a true threat to any other business.

In a conventional structure, the pyramid becomes a constraint on growth, as reward is a function of being able to fill a vacant position. In the new type of structure, growth and reward can be simultaneous. More pay is not attached to becoming a manager or waiting for a gap in the organisation. Also, there are no rules concerning pay attached to organisational position. This prevents people being rewarded for their growth in competences by having to be made a manager, a role for which they may have no aptitude or interest and none of the different competences the manager role requires.

A multi-competence team is led not managed

A conventional poor management style will have developed a number of common characteristics. These are:

- a belief that quick decisions based on minimum knowledge are a measure of successful management;
- a lack of understanding of variation in current processes and the means to achieve continuous improvement;
- firefighting (dealing with symptoms), rather than searching for root causes of problems;
- failure to co-operate across functional boundaries;
- a widely held view of subordinates as individuals who make mistakes, rather than as victims of processes that lack robustness;
- appraisal systems that reward individual performance, rather than share the rewards of improved business performance resulting from everyone's contribution to a team effort;
- a reluctance by staff to propose process improvements or highlight what appears to be their own failure in a poor process (the fear factor);
- a reluctance by staff to communicate across functional boundaries at their own level, if parochial constraints are in evidence at a higher level (the fear factor again).

Over long periods of time these behaviour patterns have become the accepted norm for managers. Subordinates take this behaviour as a role model and the pattern is perpetuated. In an ideal process, multi-competence teams would be working in robust processes without any requirement for conventional management activity. In the achievement of a lowest unit cost structure, the change from managing staff to leading teams is the way to achieve the ideal process. A fundamental requirement is the development of the role of a leader. A leader is someone who has those competences that understand processes, interact with other teams to share best practice, and act as the link to the people who are developing the DCP. The leader role is one of providing help and facilitation to the people doing the work.

To make the point, the people doing the work would appoint their own leader and appraise the individual on the basis of the leader's ability to add value to the processes in which the people work. This turns on its head the notion that a manager is there to check on whether the people have done the work. Conventional appraisals would become a thing of the past, as the leader's task is to match competences to the process and develop the competences within the team. In the new structure, we would also expect the team to recognise if the leader lacked some of the competences of leadership, and the individual would be helped to obtain the necessary skills.

The new role for leaders is largely the opposite of the poor behaviours described above. In addition, the leader will be:

- involving the people that work in the processes, because they have greater knowledge of the detail;
- co-operating across the multi-competence groups to improve the effectiveness of the whole process;
- understanding process performance through measurement;
- measuring the delivery of the DCP to facilitate its implementation;
- feeding back the concerns of current customers;
- encouraging the increase in variety and levels of competence within the group;
- ensuring that the team is motivated through recognition of its collective efforts to grow the business;
- communicating through the deployment process and participating in 'removing boulders from the runway of implementation';
- practising continual self-improvement in leadership competences.

Empowerment has been the casualty of previous management styles. Empowerment has been prevented by jobs with little perceived meaning, rigid job definitions that allow little flexibility or exercise of initiative, constant work pressure, lack of communication, uncaring attitudes, and lack of listening. Most of these problems are a reflection of the hierarchical relationships. Empowerment takes time. Instead of competing with each other, teams co-operate and count on each other to get a job done. Most importantly, staff will be accorded respect for their ability to think, challenge and innovate. At this point, the staff handbook can be reduced to one page with the single rule – 'use your good judgement in all situations'.

Building on competences

In a competency-based business, career paths can take a number of forms, as shown in Figure 6.3. Some typical paths would be:

Path 1: An individual grows in levels within a competence. Although staying within one primary skill, training and education increase the level of ability.

Path 2: An individual reaches the highest level within a competence then, through additional training, is able to become a leader of a group of people with similar competences.

Path 3: An individual changes competence but mainly uses just one of them. This would come from job rotation which would help people understand the cause and effect of their work on other groups within a process.

Path 4: An individual adds additional competences and uses all of them. Such a person would be adding value in a multi-competence team, participating in 'developing future positioning' or 'building new capability'.

Path 5: This is the path to the top jobs in the company. At this point, an individual is able to lead two types of team:

Positioning: Multi-competency teams which have a key role in developing the future positioning of the company. Chapter 5 covered the point where functions should not be working on functional elements of positioning and then acting unilaterally. The purpose of this team is to obtain the data and present it ready for conclusions to be drawn that impact on capability (the evolving DCP).

Capability: Multi-competency teams which have the key role of building the new capability to deliver the DCP. The teams will be working on each of the key business processes.

The individuals who lead teams on positioning or capability are interchangeable, as shown in Figure 6.4. These individuals constitute the top team within the business. Their key tasks are:

- Drawing conclusions from the positioning work, developing the preliminary plans, leading the deployment and feedback process, agreeing the revised plan, and initiating capability building.
- Reacting to the 'P3 alarm bell' or 'blowing the C3 whistle'.
- Optimising the use of resources between working on developing future positioning and building new capability.

This top team works together as the number one key business process and is therefore ultimately responsible for creating the company's DCP and constantly evolving it to ensure the business is growing market share and the size of the market. This is how the company will enjoy having the most adaptive and responsive mechanism to meet any emerging threat and still achieve lowest unit cost.

Maximising the value added by the top team

This top team may look very similar in composition to the current executive group of executive directors and general managers in conventional companies.

Career / Salary	Increasing added value and remuneration →
Path 1	Grow in one competence.
Path 2	L Grow in one competence, then add leader skills.
Path 3	Change competences.
Path 4	Add competences and use them all. Participate in multi-competency teams one developing future positioning or building new capability.
Path 5	P3 — Lead multi-competency teams to develop future positioning. Two roles interchange. C3 — Lead multi-competency teams to develop new capability.

Fig. 6.3 Career paths built on competences

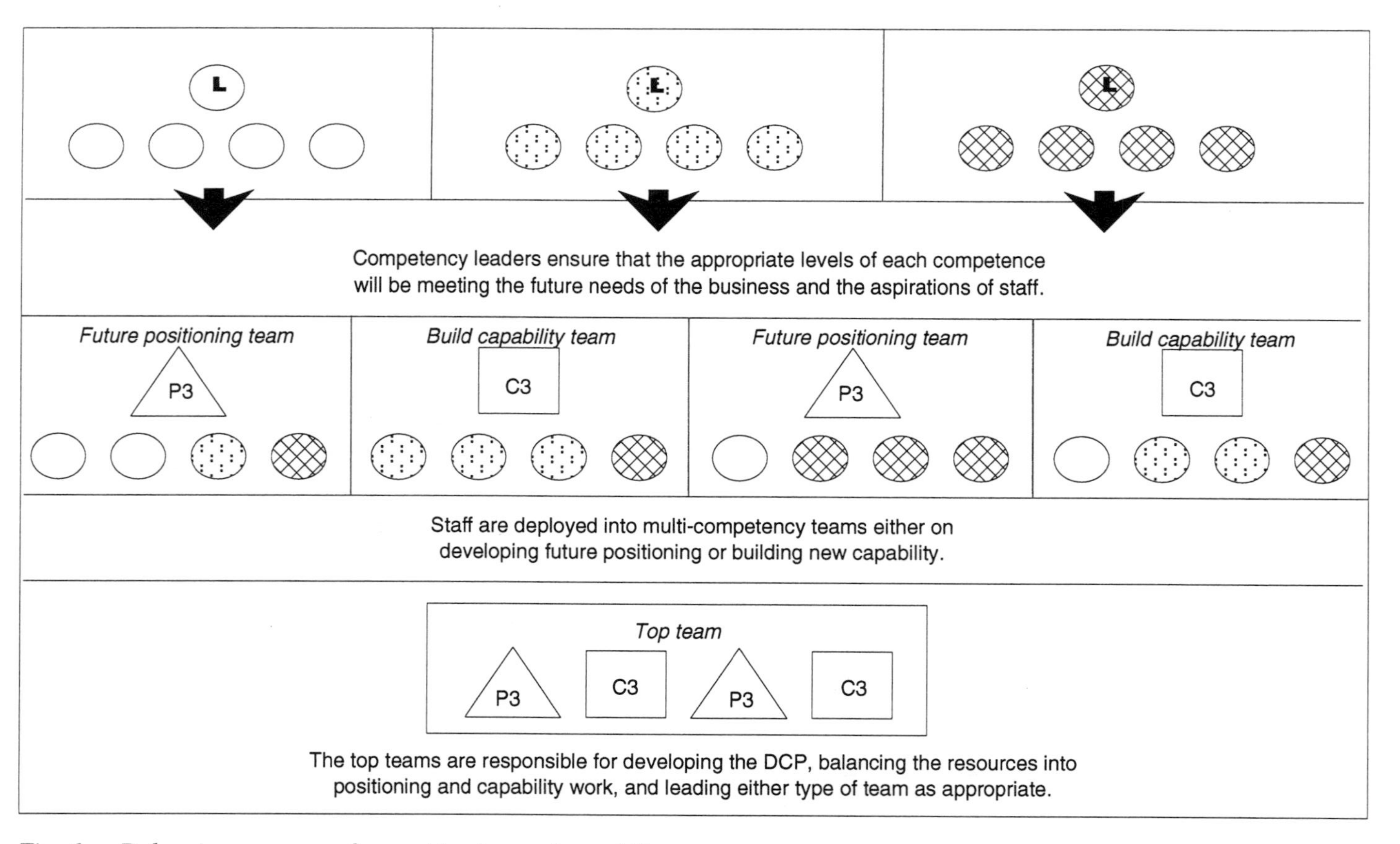

Fig. 6.4 Balancing resources for positioning and capability

However, in this structure the members of the top team have earned that role through following the competency paths to enable them to lead teams of multi-competences to work simultaneously on positioning and capability. Their rewards are commensurate with the responsibility they hold – the very future of the business.

To maximise their ability to carry out this task and to focus everyone on working in processes rather than reverting to functions, the organisation will now be able to take the structure to its ultimate point. Members of the top team need not have anyone reporting directly to them.

The business will be in control because of its approach to determining the future and deploying this to everyone so that the whole organisation is involved, setting the course and making the journey. In the processes that directly impact customers, multi-competence teams will have leaders with a new role. In the centre, groups of similar competences will have leaders with the express role of developing the groups' competences to meet the emerging needs of the business and will not be determining the use of these resources in a unilateral and functional manner. The groups of resources will be allocated flexibly to multi-competence teams, either developing the future positioning or building new capability. In either case, the groups are led by members of the top team. The top team itself will be freed from the need to achieve status or power by the old rules or be tempted to build functional towers guarded by functional parochialism.

At the start of the journey, the DCP, the point P3/C3 on the map looked attractive. As this point is approached then P3+/C3+ looks even better, and the route to it easy to take. The role of the top team is the constant management of the journey, as shown in Figure 6.5. And so it will go on, the never ending journey to being world class, and staying there, head and shoulders permanently above the rest.

BUILDING AN ORGANISATION ON THE LOWEST UNIT COST PRINCIPLE

The barrier to making radical and innovatory changes to the internal structure of a business is the fact that in the majority of cases, the current organisation is in place, operating in a similar way for many years, populated with managers and staff who have worked with each other, or otherwise, in processes that have their daily share of problems. In other words, everyone will have got used to business-as-usual however good or bad it is serving the business. For many people, this is their working life and few other options seem possible, and of those, even fewer seem likely to happen.

The principle of lowest unit cost is simpler to address by starting with a clean sheet of paper and re-building the business on a green-field site. The only conditions to apply are that as well as lowest unit cost, the business must be creating a proposition the customers will advocate. The starting criteria is thus

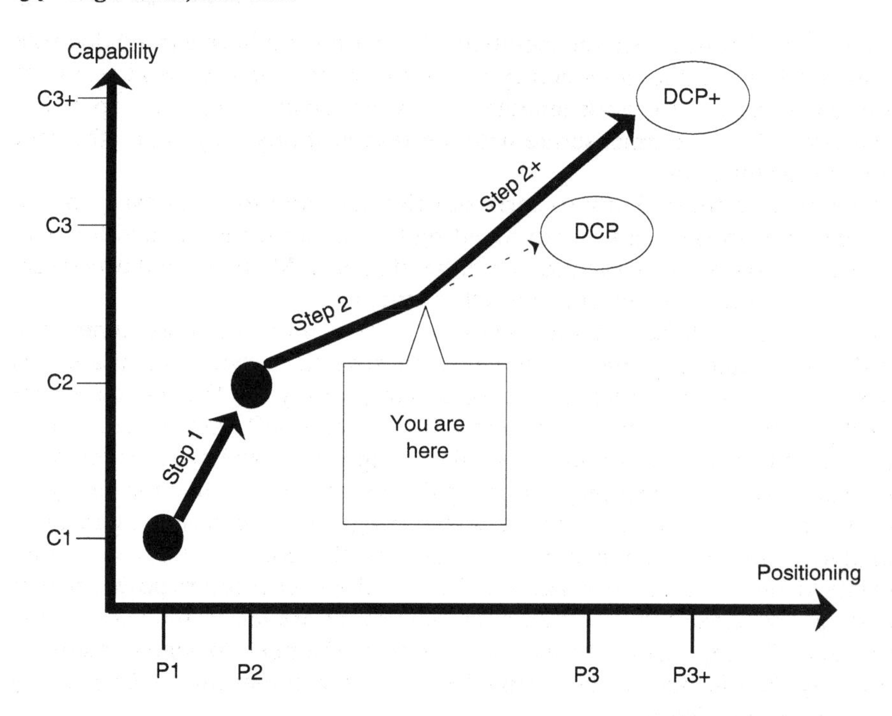

Fig. 6.5 The top team manages the never ending journey

the conclusions from the positioning work and a starting pool of people with the appropriate competences, tools and methods to do the tasks required of them. From this starting point, any business can then build up a new internal framework. To illustrate the process, we will re-build a business on a green-field site.

An example of re-building a business

In this example, let us imagine that the company completed the positioning work and found that:

- legislation had allowed many players into its sector to compete for business that was traditionally its preserve;
- players outside of its sector were encroaching on its customers and it could not respond by competing with products and services supplied by the newcomers;
- it was only represented in parts of the country and thus lost out to competitor growth;
- its customers were now influenced by many new factors that determine the nation's behaviour and these factors had not been experienced before;
- treating customers as assets to be milked only provided short-term profit and

this strategy was now detracting from creating a lifetime relationship with customers;
- targeting staff to achieve volume sales and cross-sells was alienating customers and leading to bad public relations;
- its market share and the size of the market were both declining;
- predatory mergers and acquisitions were forcing a shake-out in the industry;
- the City was still using short-term measures as a way to indicate performance, rather than recognising new customer trends or promoting long-term stability and growth as the goal.

Let us further imagine that the company also looked at its current capability and found that:

- although actions to improve current processes to meet current customer needs had lowered costs (Step 1, to move to P2/C2), much of the bureaucracy and functional parochialism could easily return again to be a barrier to continuous improvement;
- measures of staff performance and the appraisal methods were alienating staff from management and creating adverse behaviour at the customer interface;
- unit costs still made the business vulnerable to competitor discount action;
- customers were using access methods as a factor in their choice of supplier. The business could not continue to rely on its traditional delivery and distribution channels that forced its customers to act at times and at places of the company's choosing.

At this point, you may find the two lists above a useful checklist to establish whether similar conditions exist in your own organisation and to what extent they are perceived as a current problem.

The re-engineering team

The company had completed Step 1 of the BPM journey, and now members of the facilitation team continued as a group in support of short-term implementation, and trainers in the tools and techniques of continuous improvement. To address the changes to capability, C3, that the positioning conclusions, P3, had indicated were necessary, the executive deployed the need and reasons for change throughout the business. A re-engineering team, constituted from senior representatives from each major function, locked themselves in a room with the blueprint for the future, their task being to add flesh to the bones of the blueprint and detail how the internal framework could deliver the proposition that the customer would advocate, all at lowest unit cost. The team deliberated for three months, designing options, testing ideas, challenging accepted norms, and proposing innovative breakthroughs.

The logic of its thinking can be applied anywhere. The example illustrated here is a typical solution. The principles are generic and can be applied to any organisation.

The team approached the re-building task by finding answers to a series of questions.

What does a company need to just service customers at the customer interface?

The new business starts on the basis of a green-field site; the business on the inside, the customers outside. Lowest unit cost would mean that the only staff needed would be those just processing everyday business to do with transacting the customers' current needs. This arrangement is shown in Figure 6.6. Somewhere in the country, we can imagine a local market being served by teams of people with similar competences, working together in processes with appropriate procedures and systems.

> **A fundamental principle** is that the teams of different competences are linked by the processes in which they work. No managers or supervisors are needed. Any queries within the process are resolved by the people doing the work speaking to each other.

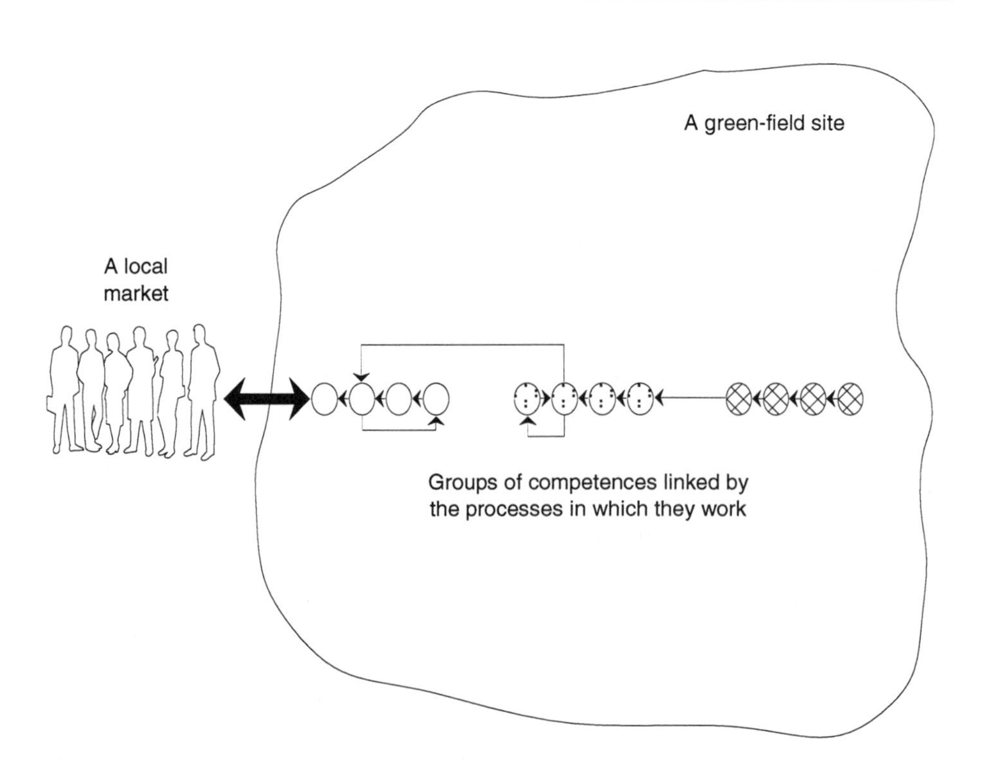

Fig. 6.6 The starting point for lowest unit cost

In a traditional company, each section of similar competences would have been reporting separately up through a function with dissimilar priorities. At this stage, no new customer needs are emerging.

How are the needs of many different local markets met, and can economies of scale be achieved?

By repeating the groups across the whole of the country, a hundred per cent coverage can be achieved. The sum of the local customer groups and the local company teams that service them can be called a 'footprint'. The key role for the footprint is to deliver the proposition the customers then advocate, through access points that the customers determine. Customers and the teams in the footprints would then be acting as partners and the competition would be locked out.

The customer-serving teams also have to change their role to be active listeners of their customers' needs and will be developing long-term relationships with customers. This brings customer retention and an ongoing need to ensure that the DCP will always be one that delights customers. It will be important to ensure that customers are advised when to buy one or a combination of 'components', rather than see every customer contact as an immediate opportunity to make an aggressive sale or cross-sell of 'products'. In a target-driven company, every sale does not always meet the customers' needs and is a cause of customers going to competitors when they are pushed off the ladder of customer loyalty.

> **A fundamental principle** is that the teams will include all the competences needed to serve the population in the footprint. As each population group served by a footprint has unique characteristics, then the mix of competences in the teams has to be different and not standardised across the country.

Usually, groups of people with different competences are needed to serve completely all of a customer's needs and, traditionally, these groups would be in different functions reporting through different lines in the organisational structure. In the footprint, these groups will all be in one team.

Where resources can be more effectively grouped to serve a number of footprints, then these can be located to achieve the economies of scale. Such groups are called 'shared resources'. This arrangement is shown in Figure 6.7. A shared resource could be located in the field, at the centre or even at home, depending on the nature of the shared resource. Shared resources are extensions of the footprints and are therefore resourced by the footprints.

> **A fundamental principle** is that a shared resource would not decide unilateral action outside of the agreement of the footprints it served. Typically, a shared resource would contain parts of the business that were initially in distinct functions, such as customer services, training, and the everyday aspects of information services.

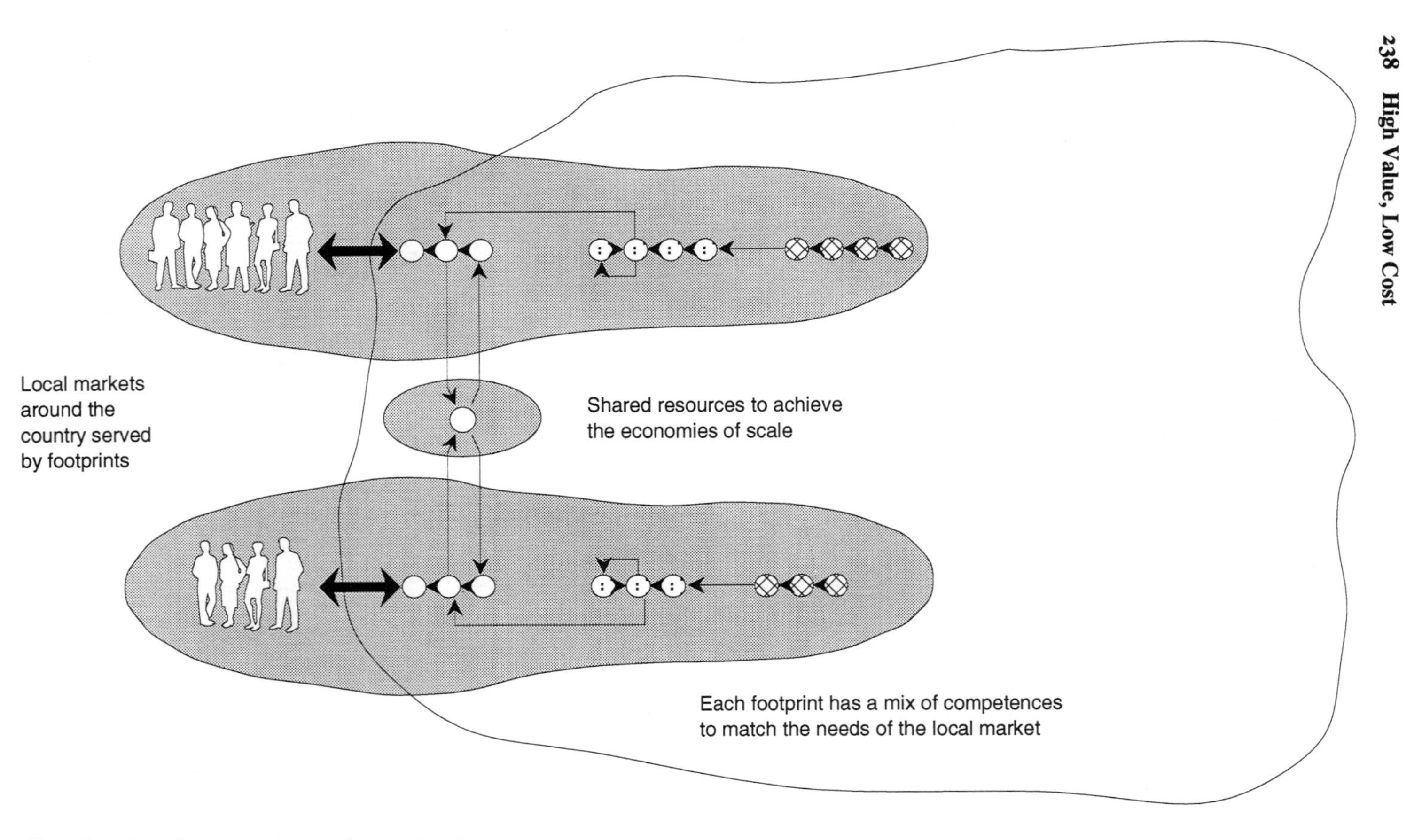

Fig. 6.7 Matching service to the needs of customers

How does the team co-ordinate in a footprint?

A prime requirement within a footprint is to achieve continuous improvement in order to reduce the unit cost of delivery. Rather than everyone co-ordinate with everyone else, the separate competency sections would co-ordinate through a leader. As the leader's role is to help the team, then the whole team elect their leader from within the team. The leader then needs a different set of competences from those used by the team. This arrangement is shown in Figure 6.8.

A fundamental principle is that leaders are people who have those competences that enable them to look at the processes within the footprint, and interact with other footprints to spread best practice. The leader's role is one of providing help and facilitation to the people doing the work.

In theory, the people doing the work could appraise the leader on the basis of the leader's ability to add value to the processes in which the people work. However, if the leader lacked the competences of leadership then the leader would be helped to obtain the necessary skills.

Because the leader facilitates local improvements, then unit costs decrease and the same number of people can process more volume. Because leaders adds more value than their cost, the principle of a lowest unit cost business is still retained. If the leader reverts back to some of the poor manager approaches, then processes do not improve and the team carries the manager as an additional cost. A manager in the old role of audit and control can cope with around eight people. In the new role a leader can lead fifty empowered people.

How do we bring in new tools and techniques of continuous improvement from around the world and still retain lowest unit cost?

To lower the cost base even further is to introduce continuous improvement (CI), something the people doing the work with their nominated leader can achieve up to a point. At some stage the teams would see the sense in investing in one of their number to scour the world for additional tools and techniques of improvement. This would still be a cost-effective business as the investment in taking out someone from the pool of people doing the work would be amply recovered by applying the new tools and techniques that were found in other companies or from academia. The centrally located 'continuous improvement person' would be funded by the teams but this investment is more than self-financing. Having found new techniques, these are disseminated to the teams via the leaders. This arrangement is shown in Figure 6.9.

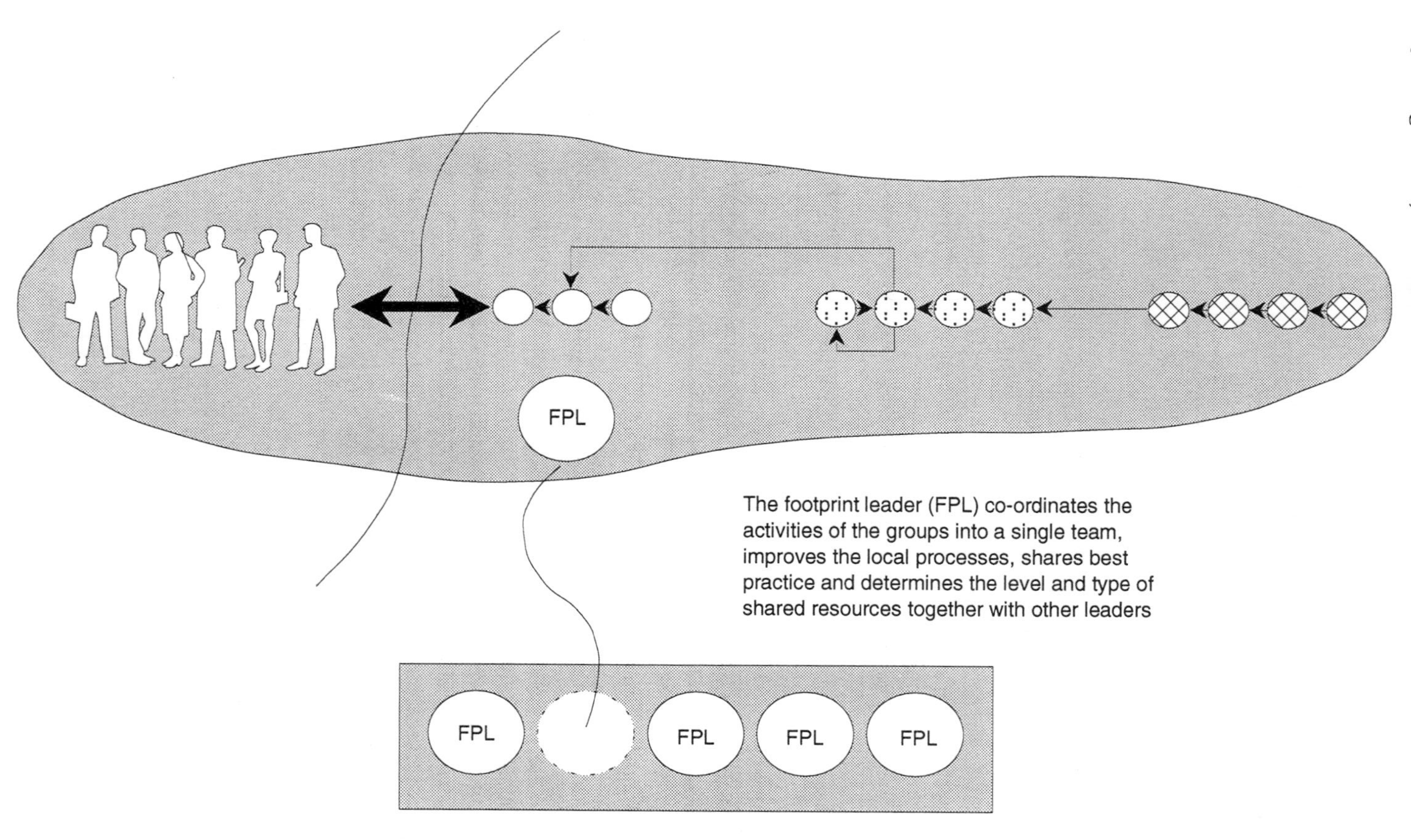

Fig. 6.8 The role of the footprint leader

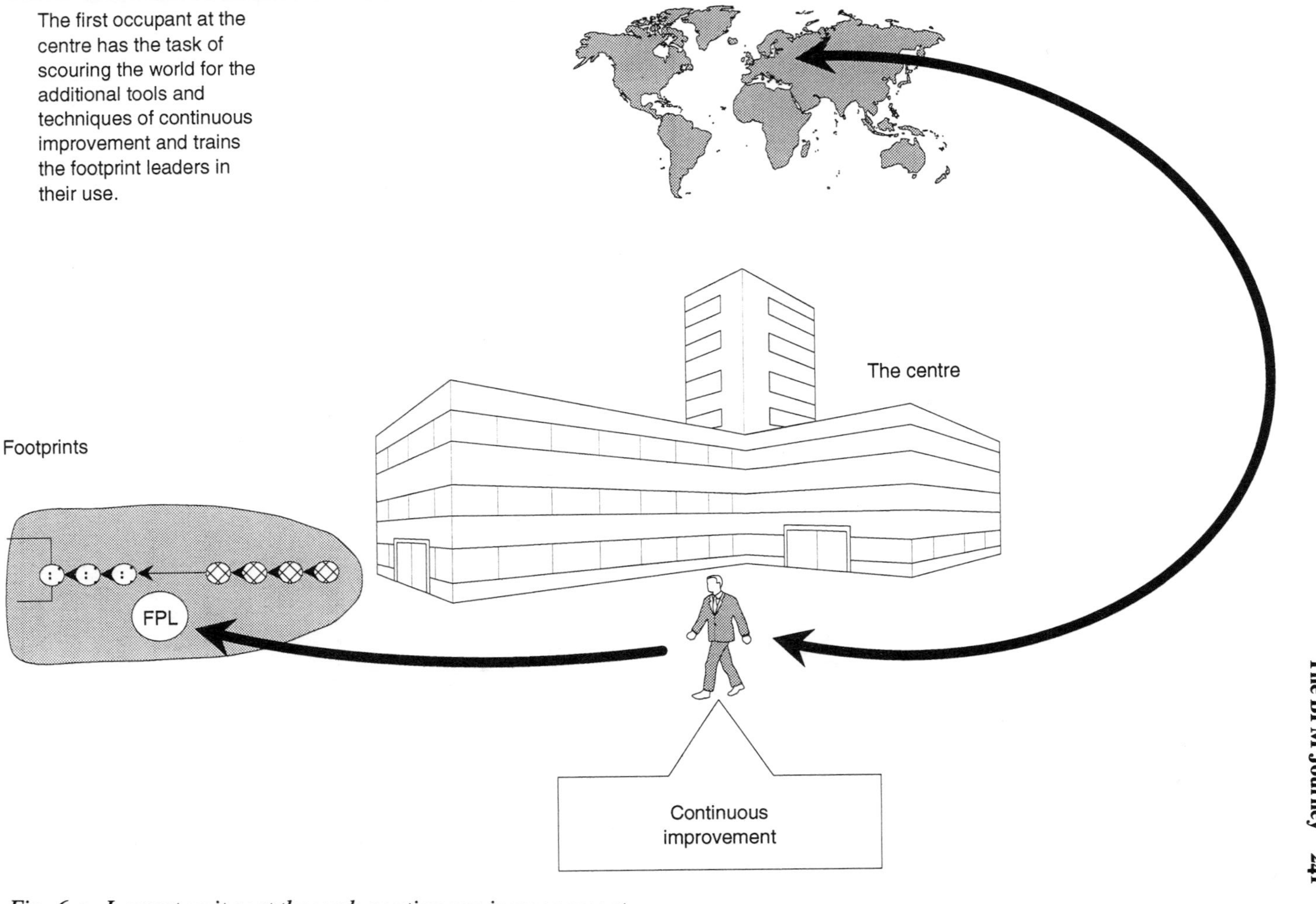

Fig. 6.9 Lowest unit cost through continuous improvement

A fundamental principle is that resources in the centre exist to help the footprints. If such resources fail to add value, then they are not required. Central resources cannot therefore grow on their own account for their own ends as this would be counter to the principle of always adding value and thus achieving lowest unit cost. This breaks the spiral of increasing costs and functional parochialism when functional heads use their authority to increase their local headcount for reasons that they do not have to justify to the organisation outside of their function.

What if the external world is changing in terms of competitor activity, longer-term customer trends, and emerging technological capabilities?

Continuously improving to better meet current customer needs at lowest unit cost will grow the business for a time. However, the external world does not stand still. If the external world is not taken into account then eventually just meeting current customer needs will not be sufficient to stay in business. The green-field company must now take this into account.

The footprints now make another investment by having some of their number given the task of:

- discovering what competitors are up to;
- discovering what the emerging customer needs and trends are in the future, and developing new innovatory approaches to delight and differentiate;
- discovering what emerging technologies can help the processes deliver at lower unit cost, and create better customer determined access points.

Quite subtly, this changes the relationship from the traditional role of the many traditional head office functions to one where the centre of influence will become the footprints and the customers. The 'developing future positioning' role of specialists in the centre brings the evidence together in order to draw conclusions on which direction to move the business. Those people in the footprints closest to meeting the customers' needs will also be in a key position to recognise when those needs are changing. In other words, both the footprints and the centre ensure that the DCP evolves. Such conclusions form the basis for developing plans. This arrangement is shown in Figure 6.10.

A fundamental principle is that no specialist will be able to action or authorise change outside the consensus of the central group working on the DCP. In the past, heads of functions have used their authority to change unilaterally key business processes without due consideration of the implications on the overall business.

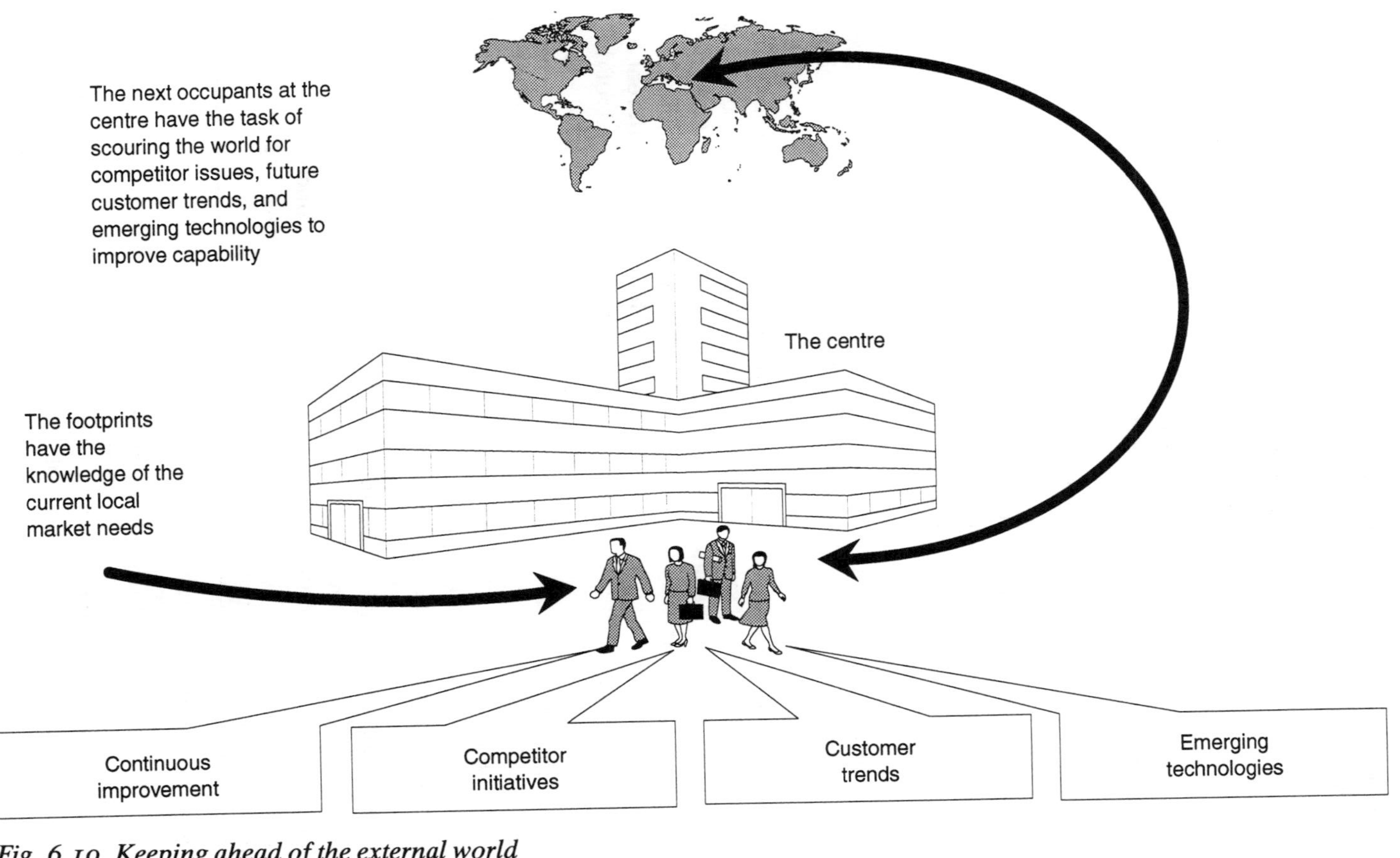

Fig. 6.10 Keeping ahead of the external world

This maintains the principle of securing the future of the footprints with the whole business self-funding at lowest unit cost.

How do we balance future positioning and future capability?

The 'developing future positioning' multi-competence experts draw on groups of people with individual appropriate competences to form multi-competence teams to look at competitors, customers and technologies. As required, the multi-competence experts change their role to one of leading multi-competence teams in order to 'build new capabilities'.

The balance between using people to build capability or develop positioning will depend on market and competitor activity, and the need to evolve the DCP to ensure increased market share and business growth. The total resources applied to both tasks will still need to satisfy the condition of achieving lowest unit cost. 'Developing future positioning' and 'building new capabilities' cannot continue with a life of their own just for interest's sake.

A fundamental principle is to ensure that the highest overall competence in the business leads the positioning and capability building teams. As both roles are undertaken by the same individuals at different times and both types of activity require a broad range of functional knowledge, there will be a risk to the business if any one individual at this level resided in one function with functional responsibility. Such people achieve the top role (the executive) at the same time as they lose organisational responsibility for a particular function. The teams they will lead perform the two different roles and the mix of competences constantly evolves to reflect the needs of the business, rather than have fixed-size groups of people in functional silos, irrespective of business needs.

The competency groups will have their own competency leaders. The competency leaders' role is to ensure that the groups' competences are matched to the needs of future positioning and capability building teams.

A fundamental principle is that the competency leader will not be able to initiate changes to previously agreed plans just because they have responsibility for a competency group. Traditionally, a functional head will often make functional process changes unilaterally because they have hierarchical authority to do so.

This overall arrangement is shown in Figure 6.11.

How are new plans deployed and what triggers a change to plans?

A fundamental principle is that the plan, created by consensus, should be deployed to the footprints who look for 'boulders on the runway' of implementation. The feedback revises the plan that can now be implemented.

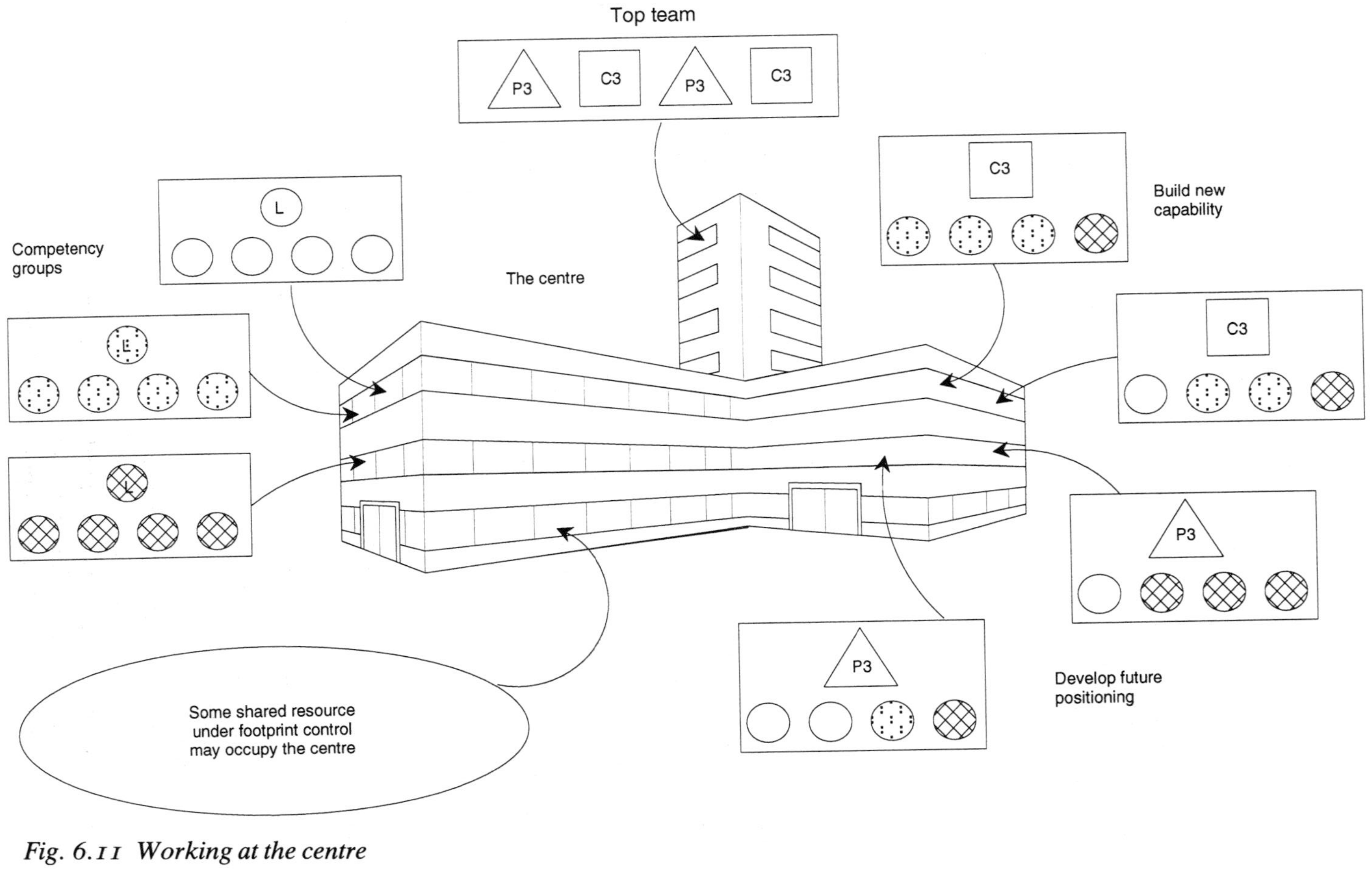

Fig. 6.11 Working at the centre

External triggers that influence positioning (customers/competitors) and internal triggers that influence capability (unsynchronised build) are considered by the multi-competence experts who head the teams that develop future positioning or build new capability. Resulting actions modify the plans at whatever stage they are at the time.

A fundamental principle is that the trigger process creates a fast response mechanism not constrained by monthly, quarterly or annual cycles. This activity we call the 'clearing house' role of the executive team. It contrasts with the traditional fixed cycle of meetings which executives reluctantly attend, weighed down by awesome agendas and reams of detailed financial printouts.

This arrangement is shown in Figure 6.12.

How would the footprints and the centre work together?

The key changes are the structure in the footprints and the structure at the centre, and the relationship between the two. Although the footprints act as autonomous units, they would be acting in a co-ordinated manner across the country to achieve a national DCP, as amended by the local DCP.

A fundamental principle is to prevent the growth of large district, area and regional offices that divorce the customer interface, dealing with current customer issues, from the central expertise, that is looking at future innovation on positioning and capability. So often regional type offices are given (or create for themselves) the role of policing the people at the customer interface using measures that are in conflict with delighting customers.

To assist the footprints, one central competency group can be devoted to helping the footprints but they should not have hierarchical ownership of the footprints. This central group has a high level of competences to help the footprints in a number of ways. Typically, this includes:

- deployment clarification;
- boulder removal clarification;
- consensus facilitation;
- feedback to centre after boulder removal;
- feedback of DCP issues at footprint level;
- members of 'developing future positioning' teams;
- members of 'new capability building' teams;
- establishing and modifying groups of shared resources.

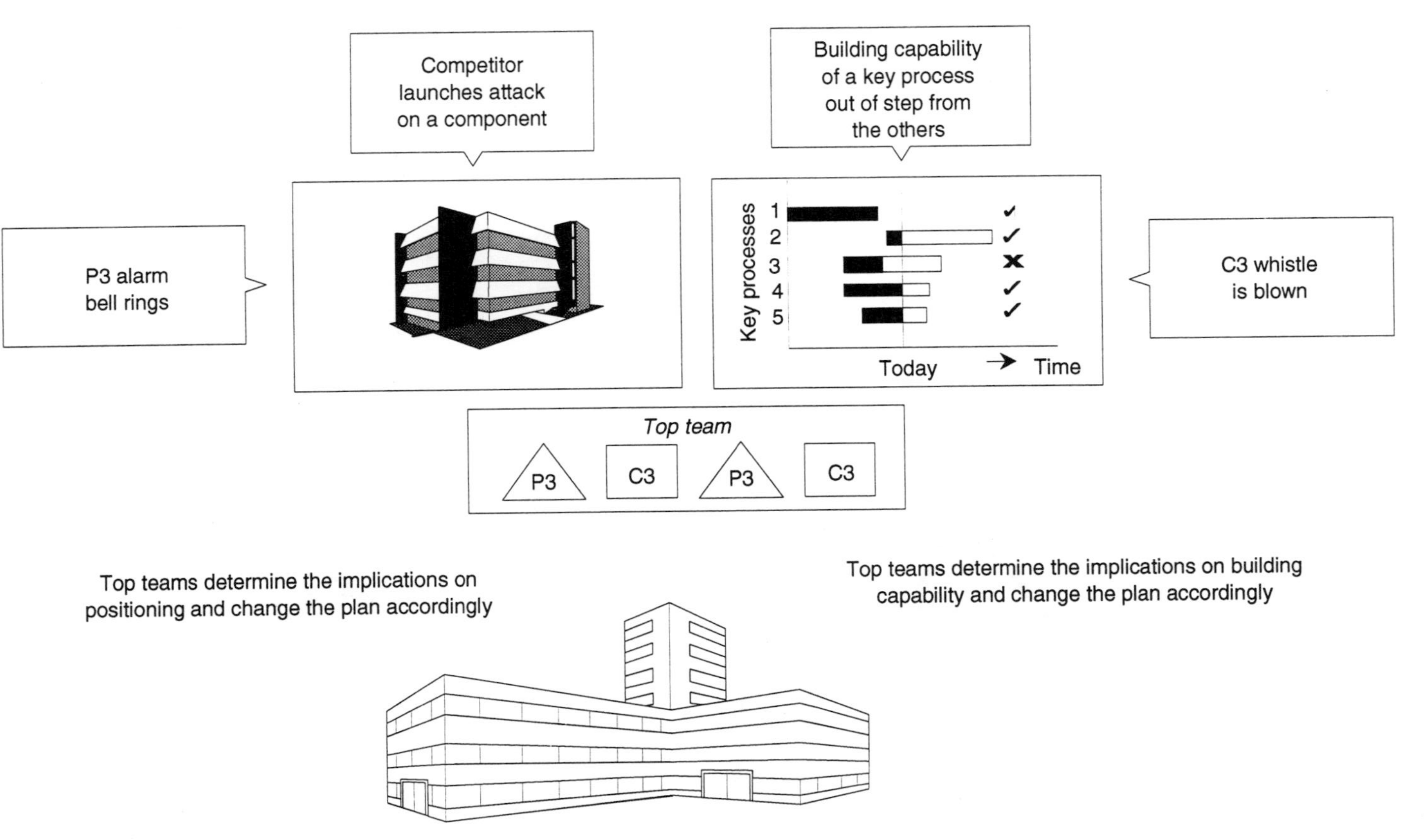

Fig. 6.12 The key role for the top team

Essentially, members of this group have individual specialist competences with which to help all footprints. A key role will be in establishing footprints and working with footprint leaders to evolve the footprints themselves as the market changes its characteristics. This arrangement is shown in Figure 6.13.

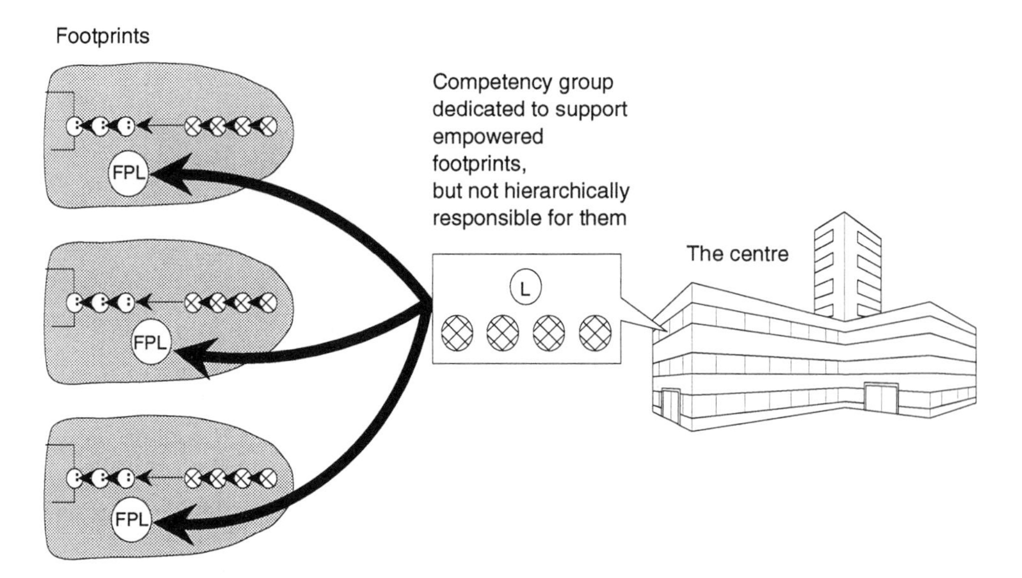

Fig. 6.13 Empowered footprints and the centre

Do any precedents exist that would give us confidence that this type of organisation can exist?

Overall, the business built on a green-field site could best be compared to a multi-celled organism, each cell or group of cells specialising in performing a role that together make the organism a living entity. Just as in nature, the organism has to balance carefully its resources and the demands of its competitor organisms and the evolving changes to its environment.

For the organism to grow and adapt, reproduce and out compete other threats to its own world, it cannot make decisions based on monthly and quarterly meetings. To optimise its use of energy and maximise its co-ordination of all its constituent parts, the legs cannot go wandering off in conflict with the needs of the body and the desires of the brain to stop and eat. But only a few centuries have passed since the industrial revolution and nature has had a number of billion years to create enviable examples of specialist adaptations of multi-cellular organisms.

THE ROLE OF INFORMATION TECHNOLOGY

The pace of development of IT in the business environment has been a combination of the following:

Vendor issues

- The pace of advances in technology from the hardware and software vendors.
- The conflict among vendors: the race to capture a world standard as a supplier set against the desire for multi-vendor integration as a customer.

Structure issues

- The retention of functionally based structures, including the IT department.
- The nature of the relationship between IT and the internal user departments.

User issues

- The uncertainties and fears of all users when faced with new technologies.
- The level of acceptance in the users' hands due to changing working practices.

Customer issues

- The nature of the cost/benefit justification for IT applications.
- The focus the business has on the outside world; transaction driven or customer relationships.

All these issues have combined to produce the characteristic uses of IT in each decade, summarised as:

1950s Mainly batch systems dedicated to discrete tasks, such as accounting and payroll. Mainly large central machines with few networks.

1960s The focus shifted to automating office manual activities wherever they existed. Resource driven with cost/benefit based on displacing staff. Mainframes with dispersed minis.

1970s Growth in word processing and electronic mail and many corporate applications. Growth in networks.

1980s Explosion in the use of powerful personal computers either standalone or networked through emulation.

1990s Distributed databases, high-speed communications, client/server environments. Higher levels of integration and platform portability. Emergence of multi-media linking computer workstation, monitor, videodisc player, card reader, printer.

All these issues have also combined to leave the impression that although looking forward, IT has always been hailed as the positive enabler of change. Looking back, each generation of change has often put the last IT initiative as a significant disabler of further progress. Although benefits in the past have been derived, the all pervading use of IT, and the dependence that businesses now have on its use, would suggest that many of the traditional views of the IT department are likely to put at risk its core position in accelerating corporate transformation.

The key now is to bring IT in from the cold and ensure that it is an integral part of all the other changes required in the business to deliver the differentiating customer proposition. For many companies, creating the proposition that customers will advocate, creating access points, and achieving a lowest unit cost internal framework, will be mainly due to the impact of IT.

Breaking the mould

The need now is to break the mould of how the IT function views, and is viewed by, the rest of the business. The IT function has often been perceived as a law to itself, doing what it likes while expressing desires to work closely with users. It has suffered from rigid budgetary control mechanisms and a desire to charge out all its cost in an effort to demonstrate its value for money or to put the total responsibility for its existence on meeting user demands. It has structured itself rigidly in sub-functions to reflect the nature of its own work patterns and a desire to demonstrate local efficiency. It has developed its own jargon that is impenetrable to an outsider and exacerbated the user's perception of elitism by locking itself away in its own buildings and paying itself large salaries. Whenever the user fraternity attempts to control and prioritise the IT function, these efforts are constantly frustrated by the actions of the IT management operating to an independent agenda.

How much of this is true and how much is myth? It is often the case that when an example of just one of these poor behaviours is observed, the rest of the business will presume the rest. It is also the case that over the years sufficient examples have arisen to re-enforce the view that the IT function will never change. However, shifting the power base for determining the use of IT to the users will not solve the problem. The answer is to develop and implement the DCP by applying the same logic to the IT function as has been applied to the rest of the business.

A role in corporate transformation

The IT function needs to break down into its constituent competences so it can play its part in the creation of the DCP and then permanently changes itself to be part of the structure that represents C3 on the BPM journey. A significant input from IT will be required in all aspects of corporate transformation:

- Developing the future positioning requires data gathering, assimilation, analysis and presentation. The use of IT in this task is a specific competence and requires careful integration with all the activities of other competences in the teams that are working on understanding and creating a proposition that customers will advocate.
- Building new capability will require IT solutions to both reducing the unit cost of the processes and in creating paperless processes with significantly reduced end-to-end process cycle times.
- The use of IT at the customer interface will enable a significant improvement

in the whole customer experience that is part of bringing the customer into partnership with the business. The integration of data and applications at a single point will enable a focus on the customer rather than a focus on the technicalities of the product.
- An interactive relationship with customers with minimum intervention from staff will decouple the customer physically from the access points determined by the business. Customers will have greater access in places of their choosing, at times convenient to them, interacting in a way that is more comfortable to them, both physically and psychologically.
- The staff themselves will be able to use IT in a way that frees them from mundane, bureaucratic, repetitive tasks to allow them to concentrate on using their expertise to service customers, and to construct and maintain customer relationships.
- Cost/benefit moves from a focus on removing staff to a focus on the value of customer retention and advocacy. IT becomes part of the holistic approach to the business and not a series of discrete deliverables to an unfocused user fraternity.

At a cursory level, one could argue that the role of the IT function has always been to provide a service that achieves the aims listed above. However, the IT function has suffered the same constraints imposed on it as all other functions in the business. Its misfortune has been that it is always a more visible target on which to vent the collective frustrations caused by the old organisational structures and practices.

The phrase 'meeting user needs' presupposes that the users are all working to the same corporate objective. Without a fully developed DCP, meeting user needs is a route to creating a whole plethora of sub-optimised solutions to problems largely caused by the lack of a DCP.

The DCP focuses the whole business and therefore gives purpose, direction and timing to all the work both on developing new positioning and building new capability. As the roles of the multi-competence teams will be well defined and the mechanisms of deployment and 'boulder removal' will be in place, both the input required from IT specialists and the needs of processes (the users) will be known, so conflicts are eliminated. Priorities are controlled by the need to balance positioning and capability activity on the large scale and the need to synchronise capability changes on the small scale.

As multi-competency teams are working on both positioning and capability, the IT expertise will not be in a separate function, divorced from the mainstream business. As with other competency groups within the business, a competency leader role will be required to ensure that the overall IT expertise is in line with future business needs. This role will not have authority to determine unilaterally how IT resources will be used.

IT's contribution to the DCP

Advances in IT developments, either from vendors or from meeting the needs of businesses, tend to become dated with great rapidity. However, the key is to drive the use of IT through considering the DCP. To this end, many recent developments in hardware, software and integration have begun to remove some of the potential constraints that would appear during corporate transformation. In some cases, the developments have provided the impetus for companies to consider how to extract DCP benefits from the available technology.

The key applications for IT are essentially:

- the customer understanding and learning process;
- the customer segmentation, customer and relationship profitability modelling processes;
- the customer experience of component building process;
- the customer access, enquiry, transaction, query, and buying processes;
- the storage and retrieval processes that support the above.

Whatever the hardware and software configurations the end result should always be a set of integrated processes that enable the business to identify and focus on profitable customer opportunities, to construct and maintain customer relationships, and to create permanent differentiation. Parts of this jigsaw can now be commonly experienced in various sectors. The whole picture can be experienced in companies that have a fully operational DCP. These are still a rarity.

Customer interactive access points

Consumers have become used to the idea of self-service, particularly were being served by others has often been characterised by the consumer being made to feel privileged and somehow grateful for having had to wait for indifferent and offhand service in locations they resent having to go to.

Self-service decouples the customer from some of the vagaries of the seller and in some cases the vagaries of other customers. Standing at the end of a queue of 'awkward' customers always seems to reflect on the supplier. Self-service can put the company into locations where the customers can be found without making a heavy investment in premises. It can also reduce the total amount of staff activity at the customer interface and where used subtly, it can stream the needs of customers and thus be perceived as better meeting their needs at any one location. This is particularly the case where transaction activity has been mixed in with giving advice and guidance.

Typically, in the finance sector, electronic and staffless branches can provide many of the needs of customers on a daily basis. Branches can be transformed into efficient, cost-effective sales and service points, rather than being community meeting places with long snaking queues. For those people requiring the type of service that needs one-to-one dialogue and explanation, more purpose-designed facilities provide the right environment where customer relationships can be formed. Where IT can provide a distributed customer database at the customer interface, then complete relationships are known at a single point and in front of the customer. A one-to-one facility includes the customer's home where staff, freed from mundane and routine transactional activity in a branch, can better concentrate on delighting customers – the lifeguiding principle. Again, IT supports the distributed customer database but in a portable form, reducing the need to access other records or make multiple visits.

Self-service touch screens and interactive multi-media are appearing in many sectors. As with anything new, customer reactions are mixed until their use becomes habitual. Some user research has highlighted some of the issues around their use. In a traditional location, self-service touch screens were heavily used by children (the 1990s child effect) and by the waiting partners of customers in a normal queue (the curiosity and boredom effect). For those people that had used the system, forty per cent made it their first choice for getting information, thirty per cent preferred talking to staff, and twenty-four per cent preferred leaflets. For non-users, the rating was fourteen, fifty-one and sixteen per cent respectively. This would suggest that the first experience of using the technology must be a significant hook if the customer expectation is low prior to use. However, as with any service, as the service across many sectors becomes high and valued by customers, any poor provider will be competitively disadvantaged.

Self-service touch screens and interactive multi-media provide more opportunities to delight and differentiate than the electronic branch. Such devices are freed from the constraint of purpose-built premises and can be located in shopping malls, airports, stations, hotels, restaurants, colleges, stores and garages. The access points can be wherever the customer determines, open twenty-four hours per day. By linking to other suppliers, a synergistic relationship develops. A 'through window' service also removes the constraint of opening hours. The system can also link commonly used services to one's own. A finance company may link the issuing of stamps, bus tickets or ski-lift passes to the system that allows the customer to buy term insurance.

The systems are not limited to just providing information. Entire sales transactions can be performed and through the use of debit and credit cards, the customer can obtain access to their own accounts to transfer cash, pay for other goods and services, update records, raise enquiries and so on. In some cases, the system can have the facility for one-way video phones that allow customers to access specialist advice from a centre of excellence and therefore reduce the need to distribute experts physically around the country.

Much research is still continuing on voice recognition. Some simple systems work on key word recognition with system prompts to obtain more detail, for example train timetables. However, voice pattern recognition and word-context recognition have not matured to a state where customers can hold any form of intelligent dialogue with a machine, even if they ever want to. Self-service, touch screens and interactive multi-media will not suit everyone but they are meeting the needs of customers who want to determine a more convenient access to a company's products and service. The machine becomes one form of lifeguide.

Although we would be dealing with a machine, research has shown that two key requirements must be met: they must be fun to use, and highly reliable. If neither of these conditions are met then customers quickly revert to taking their chances interacting with other human beings. At least they get a response if they choose to complain.

Lowest unit cost processing and building customer relationships

'Image processing', 'workflow' and 'groupware' are terms that have now crept into the vocabulary of companies who are re-engineering processes to reduce the unit cost of transactions, reduce end-to-end process times, improve document storage and retrieval activity, and improve customer relationships. The move is away from having islands of technology with discrete bases of computer power, and toward a paperless office with integrated systems at the customer interface point. Some sceptics would still point to a country-wide twenty-five per cent compound growth in paper as proof that this objective is like searching for the Holy Grail. However, significant strides have been reported by many companies. By starting from the premise that the DCP was the goal, customer service, retention and advocacy were seen as the primary reasons for making the change rather than solely office productivity. By moving away from a production line approach, one person can process the whole customer relationship, whatever the customer needs are at the time. The system is then an integral part of the lifeguiding principle which delivers substantial benefits to the customer and thus the business.

Some typical benefits that have been experienced have been immediate, rather than next day, query handling, new business processed in fifteen per cent of the time with a tenth of the people, customer retention up by twenty-two per cent, and staff turnover down from sixty to five per cent. Clearly, from the company's perspective, such benefits go straight to the bottom line. From a customer's perspective, all these improvements provided differentiation and a change in the customer relationship relative to other suppliers in the sector.

We humans seem to love paper and hoarding information. To do any task requires information from many sources: pictures, files, applications, letters, faxes, computer printouts and so on. To this we add telephone calls, disks, computer, audio and video tapes. We put things away in filing cabinets, archives, computer memories, boxes, the bin and other receptacles. Despite all this, we still forget things we should have done, throw away things we should have kept, and kept things we should have got rid of. However, to do a task means bringing these things together to make decisions and to service customers. Each time we repeat essentially the same task, slightly different conditions apply in every case. Humans are adaptive but we pay a heavy price in terms of time and errors for the way we go about things. Any 'workflow' applications have to have the same level of adaptation but need to be less cluttered in the approach and more reliable in the outcome.

In a workflow application, the data can be located in different buildings on different machines, and brings together the program's tools and data to meet a particular need at a moment in time, such as a customer enquiry. The core of a workflow application is the document management system which frees the user of the information from the type, form and location of the originals. The creation and automation of business workflows are freed from organisational and geographic constraints. Earlier in this chapter we saw how a lowest unit cost internal framework would determine when to use 'shared resources'. The technology provides the base on which a shared resource can operate without being physically co-located. In the extreme, home-based staff can contribute as if they were on the premises.

To maximise the benefits to customers and the business, workflow should be used as part of the overall changes in capability that are driven by positioning and are, collectively, the DCP. Without this context, a workflow application can default to being just a system for holding the paper in electronic format. The application requires the integration of a number of features, including the following:

- Document imaging and indexing to create a storage and retrieval process and a cross-reference to other documents or processes via the index. The image in this case is not just a replica of the original, but also the information that the original carries – its text, numbers, characters and so on.
- Through a knowledge of the business process and the context in which someone is calling up the document, only the relevant data need be presented to complete a stage in the process.
- Where the process is sequential across a number of staff, the system provides a picture of the overall flow, checking for backlogs, bottlenecks and queries raised by the staff.
- Manual procedures are prone to error when they lack robustness, leading to checking and correction activity. The system itself builds up experience and root causes of problems can be found and eliminated. Rules for error detection can be built in, freeing staff and managers from this activity and reducing throughput time.
- Through knowledge of the steps in a process and the information needs of the task, an initial activity by the staff can be set to trigger the prefetching of objects from multiple sources. The time that this happens can be seconds, minutes or hours before actual need. A customer letter scanned on receipt could trigger prefetching of their complete history for processing the next day. In any event, prefetching enables the response time of the staff to be increased through a reduction in the time waiting for the system to decompress files, access multiple systems, and so on, a key element contributing to increased productivity.
- Traditionally, computerisation has been seen as a threat to many office-based processes where it has just displaced staff. In a workflow environment

the opportunity exists to build more of the task together at one point, either at a single individual or a small team of mixed competences. In either case, the pattern of work engenders a completeness to the task that increases job satisfaction. This mirrors the similar changes in manufacturing companies where they moved towards 'manufacturing cells' containing a range of machines with operators in a team completing the manufacture of a number of parts or assemblies rather than operators working on large batches machining one detail on one machine with a short-cycle time. The increases in job satisfaction are the major contributory factors in the reported improvement in the levels of staff turnover.

- Given the empowerment that workflow applications bring to staff, the traditional role of middle managers to control, audit, check and measure staff will no longer be required. The role becomes that of a leader, as described earlier in this chapter, working with staff to improve the processes in which staff work. This is an important change as the workflow system can still be hijacked by managers who have not had the new roles deployed to them as part of the development of the DCP. The system affords the opportunity to track the process and the individuals within it. Old management behaviours that use this new information to treat staff in the old ways will create a staff reaction against the system, thus losing many of the benefits the system was designed to bring.

The reduction in lead times to answer customers queries or confirm an application or to launch new components of the proposition or to assist building the components at the time of purchase – in other words the lifeguiding principle – are all supported by workflow applications. In many respects, some elements of lifeguiding can only be achieved through workflow applications to satisfy current customers' minimum expectations. At the moment, doing this at all as a specific service enhancement creates delight and differentiation, but in the future this will become just another hygiene factor. However, to provide continually a DCP in the overall holistic sense is to be able to combine the advances in IT that allow a workflow approach to be used together with the overall corporate transformation of the business.

The real challenge is to integrate the changes in capability outlined earlier in this chapter, in terms of structure, multi-competence teams, and the balancing of resources between positioning and capability issues at the highest level, together with the emerging advances in technological capabilities that have for so long remained the preserve of IT specialists cocooned in their functional silos. Only in this way can the investment in time and money give a return that will outstrip any previous IT investment. This time, however, the return on investment comes from delivering the DCP and not from any fudged and sub-optimised cost/benefit analysis at a functional level.

7

BARRIERS TO CORPORATE TRANSFORMATION

Obviously the obvious isn't obvious otherwise more people would be doing it.
Tom Peters

WE ARE ALL VICTIMS OF HISTORY

We need to believe that people want to do a good job and take pride in the work that they do. We need to believe that when people are left to their own devices, they will naturally search for a better way to do things; not to save time so they can do nothing, but for the basic pleasure of seeing a job well done and the pleasure of seeing a better way of doing things. We need to believe that, naturally, they will help each other, as to provide help on one occasion is to receive help on another. We need to believe that where people understand the risks and consequences of their actions they will behave responsibly and not take action detrimental to others and the business. We need to believe that when people are given the proper tools to do a good job and they are competent to do something, then they can be left to do it. We need to believe that if we expect more from people, then with additional tools and help to grow their competency, they will do the job expected of them. So why is it that in the work environment, so few people feel the freedom to really enjoy work? Maybe the problem starts at birth.

As infants, we spend our early years searching for answers to the question, 'Who am I?'. We have to make a niche for ourselves in a crowded world of siblings, parents and friends. We learn attention-seeking devices and reward strategies, and the use and abuse of power among our siblings and young friends. At school, we spend some time searching for answers to the question, 'How does that work?', and to the question, 'Why does that happen?'. But we also learn that the teaching process re-enforces the attention-seeking devices, reward strategies and the use and abuse of power. We learn that if we are taught badly, despite every effort to learn, we are rewarded with low marks.

At work, we start by concentrating on how things are done as the minimum requirement to do the job and stay in the job. We follow the rules of the game to get promotion. Just like the early years, we learn attention-seeking devices and reward strategies, and the use and abuse of power in the working environment. We receive promotions and with each promotion we receive re-enforcement that we display the right behaviour and that our knowledge of managing is adequate. We learn how to be noticed by competing with our peers, even where this is in conflict with generating higher overall business performance by working together. The rules tell us that our staff should not be trusted and we make them suffer continuous audit of their decisions. We learn that their performance should be tracked by measuring things that can be measured simply, despite their irrelevance to making the business more effective. We learn that they should be stretched and asked to do things beyond their competence and we make them personally accountable so they suffer the consequences of failure.

A few lucky ones reach the top of the business and that confirms that they have the one hundred per cent of the knowledge they need to run the business. By definition, those below have less knowledge. At the top, you can listen patiently to the ideas from those beneath you but you can also draw comfort from knowing that you know more, and with your authority, squash anyone taking action contrary to your opinions. With such complacency the conditions are set to make victims of the staff and victims of the customers.

In any programme of change we hear much about 'top-level commitment' and 'the need to review fundamentally the business'. On hearing these words, why is it that staff treat them with cynicism and scepticism? They have heard it all before. Staff want to put their trust in the senior team. After all, just like the senior team, the staff want job security and to be able to take pride in what they do. Staff cannot make the business go in a new direction but they are often driven to take the business in directions contrary to the long-term building of customer retention and advocacy. Despite their better judgement, the measures that drive their behaviour constantly leave them victims of lack of trust and short-term profit generation at the expense of customers.

If treating people with decency and respect was a fundamental human right, then many managements would find themselves defending themselves in the dock. Maybe it is too much to expect an appeal to treat human beings humanely as the reason to start the change to corporate transformation. By a lucky coincidence, making such a change is also the way to survive in business and grow a profitable company. When both motives drive the change to corporate transformation, then going to work and being a customer will bring rewards to everyone. But if, through complacency, we are already confident that everything can stay as it is, we have a choice. We can do nothing, and only later discover we are on a journey to obscurity.

FURTHER READING

Trying to create an environment where corporate transformation can become a reality has much to do with one's state of mind. Changing from a conventional set of beliefs to another set is not something that happens overnight. The list of further reading is not so much a series of books containing technical methods concerning re-engineering processes. Rather, it provides insights and views which together offer another way of viewing our fellow human beings.

Byam, William C. and Cox, Jeff (1991) *Zapp!, The Lightning of Empowerment*, Business Books Ltd, London.

Deming, Dr W. Edwards (1993) *A New Economics for Industry, Government, Education*, Massachusetts Institute of Technology. Center for Advanced Engineering Study, Cambridge, MA.

Kohn, Alfie (1993) *Punished by rewards*, Houghton Mifflin Company, New York, 1993.

Seddon, John (1992) *I Want You to Cheat*, Vanguard Publishing, Buckingham, 1992.

Semler, Ricardo (1993) *Maverick!*, Century, London.

INDEX